DESERT
WINGS

DESERT
WINGS

Controversy in the Idaho Desert

Niels Sparre Nokkentved

Washington State University Press
Pullman, Washington

WASHINGTON STATE UNIVERSITY

Washington State University Press
PO Box 645910
Pullman, WA 99164-5910
Phone: 800-354-7360
Fax: 509-335-8568
E-mail: wsupress@wsu.edu
Web site: www.wsu.edu/wsupress

Library of Congress Cataloging-in-Publication Data
Nokkentved, Niels Sparre, 1947–
    Desert wings : controversy in the Idaho desert / by Niels Sparre Nokkentved.
        p. cm.
    Includes index.
    ISBN 0-87422-247-8 (pbk. : alk. paper)
        1. Military reservations—Idaho—Planning—Citizen participation.
    2. Bombing and gunnery ranges—Environmental aspects—Idaho. 3. Land
    use, Rural—Political aspects—Idaho. 4. Public lands—Government policy—
    United States. 5. Pressure groups—Idaho. 6. Coalition (Social sciences)
    7. Owyhee River Canyon Wilderness (Idaho) I. Title.

UB394.I2 N65 2001
333.73'615'09796—dc21   2001026865

CONTENTS

For my father, in memory.

In America, we have no land use ethic.

—Tom Kovalicky, former supervisor
Nez Perce National Forest

I have seen heart-stopping displays of bighorn sheep mating on narrow ledges, hundreds of feet above the canyon floor. I have been surrounded by inquisitive pronghorns, their tawny coats profiled against the black basalt. Sitting by a sagebrush fire watching impossible sunsets bathe this surreal landscape in hues of yellow, gold, and orange, sliced by the purple shadows of buttes and mesas, I have felt that I was the only human being on Earth.

The Owyhee has never been an easy land. With searing heat, Arctic cold, rattlesnakes and the distinct possibility of getting truly lost, the Owyhee does not give of itself easily. But when it reveals itself in a sudden waterfall or cougar tracks circling your camp, or a 1,000-foot chasm appearing out of nowhere, you realize that its beauty is unsurpassed. To come to accept the Owyhee on its own terms is to learn something infinitely valuable about yourself.

—Idaho attorney Brad Purdy

Preface

SINCE 1984 THE U.S. AIR FORCE had wanted to expand its training facilities in southern Idaho. Mostly the talk was internal. Then in 1989, the issue exploded in the media with the revelation of Air Force plans to take over 1.5 million acres of public land used for livestock grazing and recreation. The Air Force wanted to build a bombing range for live bombs and missiles and for low-level supersonic operations. It would have been part of one of the largest military land grabs of public lands since World War II. The public was outraged, and the proposal was defeated.

This book grew out of my reporting on the issue over those years. The story relies on my own experiences and the experiences and help of many people but in particular Herb Meyr and Bob Stevens, both former military pilots who now spend much of their time hunting and fishing and the rest of their time planning to go hunting and fishing. Both of them missed more than a few such trips when they joined the fight against the Air Force. The story also relies on a few sources who must remain nameless, but their information was used only where it squared with or supported other sources.

The bulk of the material in this book came from my own reporting. Most sources are noted, except where the source was my own reporting. My stories can be found in the Twin Falls *Times-News* from 1989 through 1999.

Many of the documents cited in this book were obtained through Freedom of Information Act requests of the Air Force at Langley, Virginia; the Pentagon; and Mountain Home Air Force Base and of the Secretary of Defense and the Defense Department at the Pentagon. I also obtained documents from the Bureau of Land Management in Boise, Idaho, and Washington, D.C., and through state public records requests of the Idaho governor's office and the Department of Fish and Game. Some of the information came from court records. In addition, private organizations sent me copies of their papers, and the Snake River Alliance and Bob Stevens turned over their collections of documents—many of which were duplicates. Copies of all records cited are in the author's possession.

I owe thanks to many people in preparing this document, too many to name them all. But I owe special thanks to Michael Frome for his inspiration and encouragement, to Kevin Richert and Gregg McNamee for their help in editing, and to my wife for her patience.

Introduction

NOT LONG AGO, the southwestern corner of Idaho was a land that few people knew. In addition to the American Indians, a few buckaroos, herders, hunters, prospectors, wanderers, and fugitives from the law came to know this raw landscape and find refuge in its wildness. Somebody in the Pentagon picked the region because there was nothing on the map—no roads, no towns. To Air Force leaders, it was like many other public lands in the West—wide-open and perfect for a bombing range. Here, Pentagon officials thought, lived few to object.

Such empty areas are often selected as dump sites or military training grounds because they are sparsely populated and offer little political resistance—not always because they are the best places. The few people living in such areas often do not have the political power, money, or influence to resist. One of the best examples of this involves the disposal of nuclear waste. Most scientists who support deep geological disposal agree that deep in the ancient granite of the eastern United States would be one of the best places to bury highly radioactive waste. But that would be nearly impossible politically. Instead, federal officials picked a site in sparsely populated Nevada, one of the weakest states politically and one where the federal government already had a strong nuclear presence—testing nuclear bombs since the 1940s, above and below the ground.

But the Pentagon officials who thought southwestern Idaho would offer little opposition to a bombing range were wrong. Joining the Shoshone-Paiute tribes in objecting to the series of range proposals were a surprising number of Idaho residents representing a variety of interests. Organized opposition grew from a historic handshake between a staunch desert preservationist and one of the leading cattlemen in southwestern Idaho at the time—two men more accustomed to opposing each other on the issue of livestock grazing on public lands in the West. The link those two forged quickly grew into a loose-knit coalition of groups and individuals that also included hunters, hikers, river runners, fishermen, archaeologists, environmentalists, ranchers, cowboys, nature lovers, and birdwatchers. These people were not against the Air Force or the military. They opposed proposals that threatened the land they cared about.

"You should go talk to some of these guys. There's the ranchers in their go-to-town outfits, next to the Eddie Bauer backpackers, next to the naturalists and archeologists in their beards and tweeds, and the Indians and the sportsmen—they're guys who like to go hunting," Idaho state archaeologist Tom Green told the *Boise Magazine*. "These people have always met to fight and argue, and now they're telling the Air Force, who wants to talk about compromise, that there's nothing to compromise about. The first few meetings one rancher just went around saying, 'Does anybody want this thing? Let's just tell 'em to go home.'"[1]

The first proposal was a massive expansion of the Air Force's 109,000-acre Saylor Creek Bombing Range northeast of the Shoshone-Paiute's Duck Valley Indian Reservation on the Idaho-Nevada border. The expanded range would have covered about 1.5 million acres on both sides of the Bruneau and Jarbidge river canyons that run north out of Nevada. Military airspace over southern Idaho, eastern Oregon, and northern Nevada already included the airspace over the reservation.

Idaho Governor Cecil Andrus favored expansion because it would make the Air Force less likely to leave the state, an insurance policy against losing forty-five hundred jobs and a $300 million annual payroll, he said. But in 1989 when the Air Force made public plans that included live bombing ranges and low-level supersonic flight, public opposition flared and political support evaporated. The proposal ran into stiff opposition that included powerful ranching interests—opposition the Air Force clearly had not expected.

The unlikely coalition of opponents was bucking a long-standing tradition in Idaho and the West. Money and influence customarily won such land use battles, and money and influence were pushing the bombing range. The proposal came during a Democratic governor's term and a Republican presidential administration with support of the state's mostly Republican congressional delegation. The range seemed a done deal. Air Force officers relied on arguments of national defense and veiled threats to close the Mountain Home base located to the north. And the Air Force was a respected agency with virtually unlimited resources and influence in Congress.

But even against those odds, the coalition of opponents defeated the proposal in late 1990. In its place, Governor Andrus proposed that the state, through land exchanges with the federal Bureau of Land Management, consolidate enough state land to create a 160,000-acre bombing range between Battle and Deep creeks north of the Owyhee River's East Fork and lease it to the Air Force. The proposal was later changed to a two-part range that would straddle the East Fork of the Owyhee River—northwest of the reservation.

The coalition of opponents eventually defeated that proposal as well. But it was replaced by a third proposal in nearly the same location as the failed Saylor Creek range expansion. This time, however, the Air Force approached the main opponent groups individually, undermining organized opposition. The Indians settled their legal challenges separately. Public land needed for the range was reduced to affect a single rancher, who was bought off. And environmentalists were undermined politically without those allies when U.S. Sen. Dirk Kempthorne tacked a rider onto a defense-spending bill, moving the proposal through Congress in late 1998—ten years since the first proposal was introduced in Idaho.

Legal challenges that arose from the earlier proposals eventually were settled out of court. Battle-worn environmentalists won a conciliatory victory, gaining some restrictions on Air Force operations, but the Air Force finally got a new training complex with practice and simulated bombing ranges and an electronic combat range in southern Idaho—without ever having shown a need for it.

Part One:

The
Phantom
Range

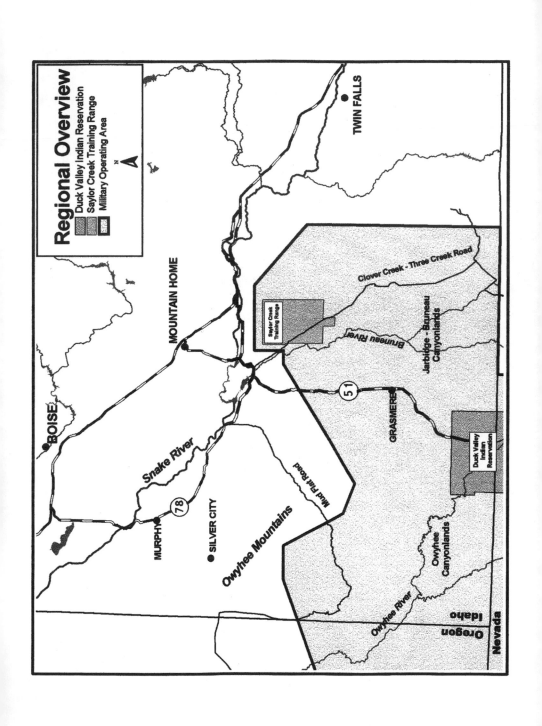

Chapter 1 / Vision Quest

KNEELING AT THE EDGE of Battle Creek, Hugo Kelly sifted the rough sand through his fingers. He was looking for special stones to use in a ceremonial rattle. The creek flows cool and clear out of the southwestern Idaho sagebrush desert. Before twisting out of sight into a steep-walled canyon of ocher lava-rock, its narrow valley flattens out for about a quarter mile, perhaps a hundred yards wide where a rough track crosses the creek. The gin-clear water, lush grass, and willow thickets along the creek offer reprieve from the harsh desert, forming a natural campsite, an oasis where generations of American Indians had camped. They hunted deer, sage grouse, groundhogs, and rabbits, gathered plants for traditional tribal medicine. Kelly, a full-blood Paiute, had come there many times to hunt, to seek spiritual guidance, and to participate in religious ceremonies.

"It feels good here," he said. A handful of sand ran out through his fingers.

In the deep pools of the creek, children pursued wily trout. Several horses were tied in the willows along the creek. Later in the day, some of those willows would be cut to form the framework of a ceremonial sweat lodge.

On this weekend in September 1993, a group of Shoshone-Paiutes had gathered to seek the help of the Great Spirit and of the spirits of their ancestors in their long-running fight against a proposal that could shatter the tranquility of the area and disturb their religious ceremonies. Since the late 1980s, they had fought proposals to expand U.S. Air Force training facilities in Owyhee County of southwestern Idaho. They feared a new or expanded bombing range would send more practicing jets over the reservation, disrupting ceremonies and disturbing life on the reservation.

Seeking relief from a hot afternoon sun in the shade of a large clump of willows, Hugo Kelly and his fellow tribal members shared their concerns over the possible effects of the Air Force's range expansion. Nearby, pots sat on a smoky fire built against the base of a rock outcrop, the pungent smells of burning sagebrush and cooking meat filling the air. High in the sky above, two fighter jets twisted and turned in mock battle. Too far away to be intrusive, the jets were an omen of what was to come.

Throughout the ten-year struggle, part of what drove the opposition to expansion proposals in Idaho was the same apparent lack of accountability that also drove opposition to military proposals elsewhere in the country. In 1989, the 1.5-million-acre Saylor Creek expansion was the largest of the military's similar proposals across the country that together made up the Pentagon's biggest land expansion since World War II.

During a time of huge budget deficits, and without assessing training needs, Pentagon officials proposed expanded and new facilities that totaled 4.5 million acres—a twenty-percent increase. The military already owned or controlled 27 million acres—an area larger than Ohio—in the form of bases and training areas that varied in size from the 3-million-acre Nellis Air Force Base in Nevada to sites of less than 1 acre. The Air Force controlled the most land with 12.3 million acres, followed by the Army with 11.2 million, the Navy with 2.3 million, and the Marine Corps with 1.1 million.

In the air, expansion proposals during the 1990s would have added 200,000 square miles of military airspace across the country. And already, more than half the airspace over the country was reserved in a variety of ways for military use. In the West, airspace expansion proposals (excluding Alaska) totaled about 50,000 square miles of new, expanded, or modified air space to accommodate more training. In Alaska, the military sought an additional 70,000 square miles of expanded airspace (see appendix).

Despite these expansion proposals, the Pentagon had not taken stock of existing training facilities, neither their capacities nor redundancies. While the Defense Department was trying to cut the number of bases, officials said increased training space was needed for new weapons that would travel faster and see farther. But without a comprehensive evaluation, the military would continue to pursue expansions, one state at a time, all over the country, said the military watchdog group Rural Alliance for Military Accountability in Nevada.

"The reasons for a national needs assessment are clear. Without it, the military will continue to acquire land and airspace in a piecemeal fashion without either a single, defined blueprint justifying those expansions, or congressional oversight of that blueprint. Without a national needs assessment, the military will duplicate existing training capabilities located on the 27 million acres which the Department of Defense already controls. Without a national needs assessment, the DOD can quietly expand upon the 50 percent of our nation's airspace that it already has."[1]

As in Idaho, people across the country were banding together to fight these military expansion proposals in the 1990s. In Massachusetts, the Upper Cape Cod Concerned Citizens community group, organized over concern for noise and danger from the artillery range at the Massachusetts Military Reservation, sued the National Guard over plans to expand the base. The effort resulted in a military study of the health problems linked to hazardous substances released at the base. Activist Joel Feigenbaum said the group's persistence paid off. Though it might not have changed military practices, it brought an awareness and studies of the problems at the base. "We would never have gotten this far if we hadn't kept at them," he said.[2]

In central Pennsylvania, local residents opposed a plan by the Air National Guard to create a new Military Operating Area over the Susquehanna River

Valley. And in Wisconsin and Iowa, local residents fought Air National Guard proposals to expand a bombing range and add two low-level training routes. In one of the longest running battles, southern Colorado residents fought more than seven years against Air National Guard plans to expand its airspace and operations over two wilderness areas. The Guard wanted to stay out of court but expected to be sued when it issued its final plan—so it kept delaying the final plan, National Airspace Coalition Director Dale Ahlquist said of the fight that gave birth to his group.[3]

One of the reasons the coalition formed was to provide a way for grassroots groups to exchange information and benefit from each other's experiences. With more than fifty proposed military airspace expansions during a three-year period in the early 1990s, Ahlquist saw a need for such a national clearinghouse.[4]

In Idaho, the battle over range expansion in Owyhee County was not much different from such battles elsewhere, and it did not go unnoticed by other groups. Activists all over the country praised the well-organized struggle, with grassroots groups eventually reaching out to national environmental organizations, involving a broad spectrum of interests, and including a successful lawsuit, Ahlquist said. As a result of that lawsuit, the Air Force could no longer treat range expansion and airspace expansion as separate issues. That was what the Air National Guard had tried to do in Wisconsin, and the judge's ruling in Idaho may have been part of the reason why the Guard dropped plans for new low-level routes.[5]

In the early 1990s, few military expansion plans received much media attention or interest from local governments or residents. That was the way the Pentagon wanted it—keep the public out of the process, Ahlquist noted. But in the cases where the public got involved, such as in southern Idaho, the military ran into real trouble. "They are destroying the very thing they are supposed to be protecting by creating veritable war zones in some of the most peaceful, tranquil places in the county," he said.

Among those peaceful and tranquil places are the rugged river canyons known to many as the Owyhee Canyonlands, where the Air Force proposed its bombing and electronic combat ranges. The canyonlands stretch from Idaho's Bruneau-Jarbidge river canyons of eastern Owyhee County into eastern Oregon and include the canyons of the South Fork and East Fork of the Owyhee River; Battle Creek, Deep Creek, and other smaller tributary streams; and the canyons of the Jacks Creek area to the north. Wildlife biologists say these canyons include some of the best wildlife habitat in southern Idaho; others say they are among the most spectacular in the country. Outdoor writer Ted Trueblood dubbed the area, "The Big Quiet."[6]

It was not always so. Owyhee County has a violent geologic history that belies the tranquility one can find there. According to one scientific theory, approximately seventeen million years ago a meteorite slammed into what is now southeastern Oregon. It cut into the earth's upper mantle—the layer of hot rock beneath the continental crust. Lava flowed from the resulting crater, forming what is now the Columbia Plateau and the Owyhee Plateau in southwestern Idaho.[7] Ancient lava flows dammed up an enormous lake, Lake Idaho, that began to drain once the water had cut Hells Canyon of the Snake River. As the lake drained, other streams and rivers cut deeper into the plateau in southwestern Idaho. Erosion worked backward, excavating canyons to their present depths. A similar situation can be seen today along the Snake River, where Shoshone and Twin falls represent the headward erosion and deepening of the Snake River Canyon.[8] A more famous example of similar activity is Niagara Falls in New York state, which are slowly eroding their way toward Lake Erie.

In this manner, the Owyhee River, its forks and tributaries, and the Bruneau and Jarbidge rivers have carved their steep, narrow canyons through the Owyhee Plateau. The East Fork of the Owyhee drains the Elk Run Mountains of northern Nevada and runs northwest across the Shoshone-Paiute Indian Reservation. As it crosses the reservation it spreads into a broad, wetland valley that is the breeding ground for numerous species of ducks, giving the reservation its name, Duck Valley. Farther west the East Fork joins the South Fork, which drains the Bull Mountains in northern Nevada. Together the two forks run into southeastern Oregon, where, as the Owyhee River, it eventually flows north into the Snake River. On its way, the East Fork winds through the biggest empty space in the lower forty-eight states. The Bruneau and Jarbidge rivers drain high basins in the Humboldt-Toiyabe National Forest in northern Nevada, cutting north through the plateau to the Snake River.

Since the last Ice-Age glaciers retreated from southern Idaho—and perhaps before that—the area has been home to humans. Scattered along Pole and Camas creeks north of the East Fork of the Owyhee River, prehistoric humans left abundant archaeological evidence of their habitation—rock shelters, hunting blinds, and chips from the production of stone tools. The number, size, and distribution of archeological sites make the area the most complex mosaic of prehistory in the state. Sites have been dated as far back as 6,000 years ago, but the most active settlement existed from 600 to 1200 C.E.[9]

In the early 1800s, British fur trappers became the first whites to explore southwestern Idaho. American fur trappers soon followed. In the mid-1800s, southern Owyhee County was still largely unexplored, known only as the God-forsaken country immigrants passed through along the Snake River on the harder but shorter southern alternative of the Oregon Trail. Then in the early 1860s, miners found gold in Jordan Creek in the Owyhee Mountains.[10] The gold strikes and a growing cattle industry to feed the miners brought white settlers into Owyhee County. Huge cattle herds trailed up from Texas—as many as 200,000 head roamed Owyhee County before the killing winter of 1888.[11] One of the most striking ghost towns in the West still stands as a testament to the

Confluence of East Fork and Deep Creek. *Photo by the author.*

mining activity that flourished in the Owyhee Mountains—Silver City, "Queen of the Owyhees," perched in a narrow valley at six thousand feet.[12]

Owyhee County was officially designated by the Territorial Legislature on December 31, 1863. The word "Owyhee" was an early, phonetic spelling of Hawaii, also known then as the Sandwich Islands. In 1819, three "Owyhees" working for Donald McKenzie were sent to trap a large stream that emptied into the Snake River. McKenzie was a partner in Canada's North West Company. When the three did not return, McKenzie investigated. He found one of the men murdered in camp and no signs of the other two. McKenzie thought they had been killed by Indians. The stream and the region were named in their honor. Some maps of the day also referred to the Owyhee River in eastern Oregon as the Sandwich Island River.[13]

In later years of the 20th century, the rugged canyons of the Owyhee River's South Fork sheltered the operations of outlaw poacher Claude Dallas. In January 1981, Dallas killed two game wardens—Bill Pogue and Conley Elms of the Idaho Department of Fish and Game—who had come to his camp on the South Fork to confront him over illegal trapping and poaching.[14] But cattle rustling was and still is the biggest crime in Owyhee County, and in its 7,643 square miles—about the size of Massachusetts—there are plenty of places for rustlers to hide. The county extends east to west from the Twin Falls County line to the Oregon border and north to south from the Snake River to the Nevada border. The county averages about five inches of precipitation per year—most of that as winter snow. And more cows than people live in this remote corner of Idaho. Most of the 9,500 residents of Idaho's second largest county—4.9 million acres, 82 percent of it federally owned—live within a few miles of the Snake River.

Only two paved roads traverse the county. Idaho Highway 78 runs along the Snake River from Bruneau to Homedale while Idaho Highway 51 runs south from the Snake River to the Duck Valley Indian Reservation. Two other roads— both gravel—provide most of the access to southern Owyhee County. Mud Flat Road runs south from Grandview up onto the Owyhee plateau and turns west into Oregon. The Bruneau-Three Creek Desert Road runs north to the town of Bruneau from the ranches at Three Creek in the southeastern corner of Owyhee County near the Twin Falls County line and the Nevada border.

The Bruneau River was the last of Idaho's white-water rivers to be explored. "The basaltic rocks rise perpendicularly so that it is impossible to get from the plain to the water or from the river margin to the plain," the famed explorer Capt. Benjamin Bonneville said in 1833, describing the forbidding river.[15] In the summer of 1950, three Twin Falls men, Len and Stan Miracle and Jonathan Hughes, navigated the Bruneau River Canyon in an Army surplus raft. Since then, the Bruneau and its main tributary, the Jarbidge, have become popular springtime raft trips. Both are steep, difficult rivers—as Bonneville noted, the canyon walls prevent any escape once on the river—popular with expert white-water boaters around the world. Twenty-nine miles of the Jarbidge and seventy-one miles of the Bruneau have been recommended for federal protection as wild and scenic rivers.

A good portion of other wild rivers in Owyhee County have also been rec-ommended for protection: twenty-six miles of the Owyhee River's South Fork; sixty-six miles of the East Fork; twenty-eight miles of East Fork tributaries; and eighty-eight miles of Deep Creek and its tributaries. The area that includes Big Jacks Creek, Little Jacks Creek, and Duncan Creek has been proposed as wilder-ness. It includes fifty miles of rugged, meandering canyons that are nearly undis-turbed by humans and are ideal for observing wildlife. The BLM already has classified much of the area as a Wilderness Study Area.

In southern Owyhee County, near the Nevada border, the dirt road to the Shoshone-Paiutes' camp at Battle Creek heads west into the unpaved corner of southwestern Idaho. The road—listed as a jeep trail on USGS topographical maps—leaves civilization behind at Idaho Highway 51 just a few miles north of the Duck Valley Indian Reservation. For most of the twenty miles—it can seem like fifty—to Battle Creek, the road bucks and bumps over rocks that threaten to slash tires and tear out vital underparts. (During a break in my driving to the camp in September 1993, as if on cue, a sonic boom shook the rocky landscape as an Idaho Air National Guard F-4 Phantom fighter jet outraced its own roar across the sky.)

Here, in this environmentally sensitive, popular recreation area, important to wildlife and well-used for livestock grazing, the Air Force proposed its high-tech, supersonic practice war zone. Through ten years, conservationists, hunters, Indians, hikers, river runners, farmers, and ranchers across southern Idaho would fight the Air Force's efforts. Officials used misleading and false information—outright lies at times—and manipulated the public process with the help of Idaho political leaders to get a training complex that was not needed. Air Force officials insisted publicly at times that a new range was vital to national security and that the Mountain Home Air Force Base would be closed without it. Idaho politicians said that Air Force pilots deserved the best training the country could give them, implying that without a new range in Idaho, training was somehow less than that. Meanwhile, Air Force lawyers testified under oath in federal court that a new range was not necessary to provide that training. Air Force officials asserted—no doubt correctly—that a new range would improve training for pilots in Idaho, but they never were able to show that the improvements or the range were needed.

Without the Defense Department having to assess its needs or justify its proposals, how many other military training areas across the country are wasting tax dollars and elbowing their way into the increasingly crowded wide-open spaces of the West? If Congress will not hold the military accountable, who will?

Most of the country in southwestern Idaho where the Air Force wanted to put its training complex still looks largely the way it did when Columbus landed. Native grasses, shrubs, and flowers still grow here in places where water holes are too far away for the plodding, domestic cattle that graze most public lands in the West. No fences or power lines cross the rugged, rolling grassland, dotted with sagebrush and ancient gnarled junipers and booby-trapped with sharp, broken lava and unexpected canyons. For five months of the year, weather keeps most humans out, and that makes the high desert plateau superb wildlife habitat. Here Indians still find refuge for their vision quests, a place to perform religious ceremonies without disturbance, a place where ancestors fought and are buried, a place of tranquility and reverence. And the world-weary of today find solitude and humility, a place where on most days all one can hear is the hot, dry wind and one's own heartbeat. It is a place so wild that if the unexpected happens, a person could die—a place against which, as Wallace Stegner said, to measure one's spirit.

"There's something positive here that should be preserved," said Dan Press, a Washington D.C. attorney who represented the Shoshone-Paiute tribes. To him, preserving the land for all the reasons that various groups wanted to stop the range was important—and it was important for those groups to work together, he said. Individually, they were not strong enough to stop the range; together they might be.

The Indians had fought to preserve this land once, and they would fight for it again, tribal official Lindsey Manning said. But this time they were not fighting alone.

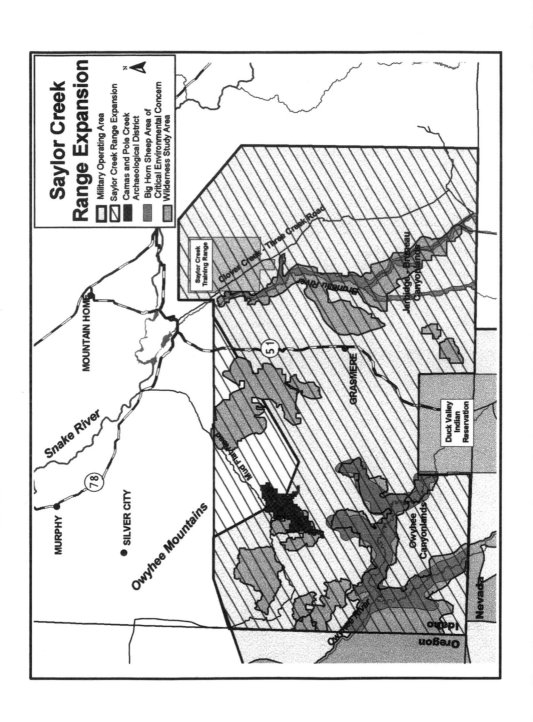

Saylor Creek
Range Expansion

- ☐ Military Operating Area
- ▨ Saylor Creek Range Expansion
- ▨ Camas and Pole Creek Archaeological District
- ▨ Big Horn Sheep Area of Critical Environmental Concern
- ▨ Wilderness Study Area

N

MURPHY

Snake River

⑦⑧

SILVER CITY

MOUNTAIN HOME

Owyhee Mountains

Saylor Creek Training Range

Clover Creek - Three Creek Road

Bruneau River

⑤①

GRASMERE

Mud Flat Road

Jarbidge - Bruneau Canyonlands

Duck Valley Indian Reservation

Owyhee Canyonlands

Owyhee River

Oregon
Idaho

Nevada

Chapter 2 / Hometown Activist

IN 1984, WHEN HE RETIRED from the Air Force as a colonel, Herb Meyr had heard talk of plans to expand training facilities at the Mountain Home Air Force Base, where he was stationed. Air Force officials wanted to enlarge the Saylor Creek Bombing Range southeast of the base and to add a few radar sites in the desert to accommodate increased training needs. Then during a visit to Idaho in 1984, an Air Force deputy assistant secretary, Gary Vest, told officials they were not thinking big enough. They should look at long-range plans and future Air Force needs for the Saylor Creek range. Vest urged them to look toward emerging weapons systems and new generations of warplanes. What eventually would become the Air Force's latest fighter jet, the F-22, was on the drawing boards at the time.

Vest was well-connected politically in Idaho, a University of Idaho classmate and fraternity brother of U.S. Sen. Larry Craig and former Sen. Steve Symms, both Idaho Republicans. Martin Peterson, another Vest classmate, was an Idaho Democrat, executive officer of the state's Centennial Commission, and past chief of the Idaho Division of Financial Management under Governor Andrus.[1]

Over the next four years, officials at the base set out to develop plans and alternatives for expanding the training facilities in Idaho, including live bombing ranges, supersonic operations, and electronic combat ranges. Tactical Air Command headquarters approved the ambitious expansion plans at Mountain Home in December 1988. Officials at the base scheduled public hearings for October 1989 and planned to have the range ready in 1994. But on December 29, the federal Commission on Base Realignment and Closure issued its decision to move ninety-four aging F-4 fighter-bombers to Mountain Home from George Air Force Base in California, which was to be closed. The decision forced the Air Force to accelerate the range expansion schedule. Instead of 1994, the range now had to be ready by 1992, Air Force officials said.

The commission may have based its recommendation on Air Force plans, but few in Idaho knew much about those plans at the time. Meyr had not heard any more since he retired. That would soon change, however, with the contributions of him and two other men, despite the efforts of the Air Force to control the release of information. A Mountain Home dentist helped to alert the environmental community and organized a meeting that brought together people who would form the basis of the opposition. An Owyhee County rancher represented the interests of ranchers at that meeting and later in front of Congress. And Meyr brought his experience and expertise to give credible critique of the Air Force proposal.

But the expansion plans, about which the Air Force had said little, became entangled in the country's effort to close unneeded bases. The Air Force tried to claim the commission's decision required the expansion.

In a budget-cutting move in May 1988, the federal government formed an independent commission to study and recommend consolidation of military bases. The Defense Secretary's utilization of the Commission on Base Realignment and Closure was an attempt to avoid the political tug-of-wars that had hampered past efforts to close unneeded military installations.

Following the Korean War and until the 1970s, Congress closed approximately two hundred bases. Following the Vietnam War, the military closed approximately five hundred more installations around the country. The process angered Congress, whose members felt the pain of bases closing in their districts. Congress failed in its effort to block the executive branch from closing bases without congressional approval, but it succeeded in requiring environmental impact statements for any base slated for closure. As a result, the effort to close unneeded bases all but stopped. The process resumed in 1988 with the nonpartisan Base Realignment and Closure Commission.[2]

The commission was composed of twelve individuals and a professional staff, more than half of whom were Air Force or other Defense Department employees. The commission's executive secretary, research director, and counsel were also Defense Department employees. The commission's job was to devise a way to identify bases to be closed or realigned, reduce any negative effects, and develop criteria for realigning and closing bases, all by November 15, 1988. The commission's final report was due by December 31. The Base Realignment and Closure Act stipulated that Congress would accept or reject as a whole the commission's recommendations in that report.

The commission explained that while cost reduction was an important consideration, the military value of a base should be the preeminent factor in any decisions.[3] Data supporting the process was provided by the military and would be validated by the commission and its staff. But much of the testimony presented to the commission was taken in secret with no public scrutiny. Opportunities for public comment came only after the military's recommendations had been lodged with the commission. The commissioners argued that, despite appearances, they were independent and did not blindly accept whatever the military recommended. The commission was not a rubber stamp. It had the power to make changes in the military's recommendations.

"The commission was established as an independent body to make recommendations to the President after analysis of a list of recommendations from the Secretary of Defense," commission Chairman Jim Courter wrote. "After review of all factors and comments from the Congress and public, where 'substantial deviation' from the selection criteria or force structure were determined, the commission independently recommended changes without consultation from any outside party. This is in accord with public law and necessary for the base closure process to continue with the confidence of the executive branch, the Congress and the general public."[4]

A skeptical Congressman Jerry Lewis of California said, "Does the truly remarkable effort by the sponsors and supporters of the Base Closure Commission statute to insulate the process from congressional pork barrel politics as usual justify a decision process which met behind closed doors? Can it stand up to the charge that the Pentagon has a vested interest in logrolling a Congress which has delegated its informational powers to a Commission totally dependent on the analysis which each branch of the service chose to give it?"[5]

In their December 1988 report, the commissioners concluded: "The recent expansion of the electronic-combat and weapons ranges in the Mountain Home area provides the capability to relocate operational and training assets, which will increase efficiency and enhance mission effectiveness." And: "The military value of George AFB is lower than other tactical-fighter installations due to its distance to specialized training ranges and the increasing air-traffic congestion in the vicinity of the base."[6]

Those conclusions were based on the testimony of Air Force officials; the commission staff apparently did not check the information. In an undated transcript of commission testimony, one of the commissioners—his name was blacked out—asked a commission staff member, "In fact, is this the Air Force's solution to closing: That's not your solution, is it?"

"They followed our process and came up with this solution. And we validated that that was good work, good analysis," replied the staff member, whose name also was blacked out.

Government documents, aeronautical charts, a public information officer at George Air Force Base, and a report by the General Accounting Office—the investigative branch of Congress—told a different story. There had been no expansion at Mountain Home or at the Saylor Creek range. Mountain Home had no low-level supersonic operating area, no live ordnance range, and its electronic combat simulators could not accommodate the F-4s' training requirements. The base had only one range, rated as marginal by the Air Force, twenty-two miles away and too close for any direct approach. In fact, less than two months after the commission's decision, Air Force officials said "significant modification to both range and airspace will be essential to realignment" at Mountain Home.[7]

Yet the commission had said: "Suitable alternative locations had to be identified for each major activity or unit before proceeding with further consideration of closing an installation."[8]

The commission's conclusions about Mountain Home appeared to be based on Air Force expansion plans already approved by the Pentagon. On February 23, 1989, Col. Timothy L. Titus, an Air Force legislative liaison officer, told the House Subcommittee on Military Construction Appropriations that "a range improvement program in existence prior to the Base Closure initiative will expand the existing ordnance impact area to approximately 900 square miles."[9] Plans also included a $31 million program to improve the electronic combat capability of the Saylor Creek range. "No additional procurement of instrumentation would be required to conduct training," Titus said. Air Force officials apparently had convinced the commission that expansion would go ahead because the Pentagon had approved plans.

But the commission report was wrong about George Air Force Base as well. Pilots from the base used the Tonopah electronic combat range, a part of the Nellis Air Force Base's range complex in Nevada, for only about twenty percent of their training. The closest access point to the range was 120 nautical miles from George. Most of the training took place at ranges that are part of the 20,000-square-mile restricted airspace complex known as R-2508, which begins just a few miles from the base. Additional operating areas in Nevada and off the California coast brought the total training airspace available to pilots at George to more than 57,000 square miles. Pilots had access to nineteen low-level routes for navigation training, six electronic multiple-function training ranges, supersonic operating areas, and ocean ranges for practicing with live air-to-air missiles.

The commission also was misinformed about air traffic congestion in the Los Angeles area. The 10,500-foot San Bernardino and San Gabriel Mountains between George and the Los Angeles Basin form a natural barrier between low-flying military jets and higher flying commercial traffic. Commercial planes into and out of Los Angeles International Airport fly above 20,000 feet while George aircraft typically fly between 7,000 and 11,000 feet.

"When is it OK to lie to Congress?" Rep. Jerry Lewis of California asked the House Armed Services Committee, Subcommittee on Military Installations and Facilities, on February 22, 1989. "It is absolutely not true, as the Commission report suggests, that the Saylor Creek target range at Mountain Home is adequate for either the pilot training or electronic targeting exercises essential to the mission of F-4 aircraft." Lewis represented the district that included George.[10]

On the House floor, Lewis said, "However well-intentioned the Commission was, its process was flawed by a dependence upon the very services whose goal was to protect what they had or at least to justify massive new military construction in the 1990s."

Herb Meyr did not hear more about the expansion plans until the summer of 1989 when over beers one day at the officer's club at the Mountain Home base, a squadron commander sketched out the proposal on a cocktail napkin. The sketch showed five target impact areas straddling the Bruneau River. Meyr was shocked at the size. The 1.5-million-acre range was a massive facility for a bunch of aging fighter jets. He had been a fighter pilot and an instructor of fighter pilots; he knew that a few squadrons of F-4s did not need all that. It would be a real bombing range.

Meyr had earned a Distinguished Flying Cross for flying into the teeth of enemy fire in the Mekong Delta of South Vietnam in 1966 to save a group of soldiers. He had grown up with airplanes. He was born in Coral Gables, Florida, shortly before World War II, son of a pilot who flew flying boats for Pan American. He became an Air Force fighter pilot in the early 1960s, and following tours in Europe and the Middle East, Meyr was sent to Vietnam. He flew 448 combat missions, logging more than six hundred hours. Meyr returned to the United States to help train Vietnamese and American fighter pilots. In 1969, he was transferred to Nellis Air Force Base where, a few years later, he flew in some of the first combat-crew training sessions known as Red Flag, the start of organized composite force training in the Air Force. In 1977, Meyr moved with his squadron of F-111s to Idaho, rising to chief of standardization and evaluation at wing headquarters—an expert on fighter pilot training. He knew only too well what was at stake for fighter pilots in combat and what was required in training.

Meyr had flown over the Owyhee County canyon country with the Air Force, and since retirement he had hiked, camped, and hunted in much of it. What he had heard of the proposal from his friend did not add up. So later in the summer when he heard about a public workshop in nearby Glenns Ferry, he decided to attend, mostly out of curiosity, he said. But also in hopes of learning more.

He eventually became one of the most effective and persistent critics of the range proposal. He was a hometown boy, one of the Air Force's own. He could not be passed off as some environmental wacko or out-of-state hired gun. But his position eventually made him unpopular in his hometown. Local businessmen tried to discredit him, and his wife would lose her job with the contractor that maintained the targets and ran the scoring equipment at the Saylor Creek range, Meyr said. But his record spoke for itself; his credibility was unimpeachable.

Meanwhile, the first public inkling of expansion plans at the Saylor Creek Bombing Range had come at a little publicized "scoping" meeting of Air Force officials and a few local residents on March 16, 1989, in Mountain Home. Col. Danny D. Howard of the Mountain Home Air Force Base explained the plans to conduct an environmental impact statement on the proposed move of ninety-four F-4s to Mountain Home. The statement would cover the significant effects of aircraft realignment and range and airspace modifications. But he said nothing of a fifteen-fold expansion of the Saylor Creek range, or of low-level supersonic flight and live bombs. He did not know the size and limits of the range, but he thought supersonic airspace would not be required to support F-4 training.

Few attended the meeting, and word of the discussion did not spread. Soon afterward, however, Mountain Home dentist Randy Morris was asked to present a slide show on the Owyhee Canyonlands to the Mountain Home Chamber of Commerce. Morris had been working on a national park proposal in southern Owyhee County for almost a decade. Along with the Committee for Idaho's

High Desert, a desert conservation group, Morris had prepared a slide show to help push the park idea. He showed up expecting the entire chamber, interested in hearing about the park proposal. Instead he found just a few chamber officers and Lt. Col. James Cooper from the Mountain Home Air Force Base, who apparently were more concerned that the park proposal should get in the way of the proposed range expansion. During the slide show someone remarked that these pictures were from out where the range expansion was proposed. Cooper just grunted. Morris's heart sank. He had fought the ranchers, the BLM, motorcyclists, and miners on the national park idea. Now, it seemed, he would have to fight the Air Force, too.

But the full impact of what the Air Force had in mind did not hit Morris until he attended a June 28 meeting that U.S. Sen. Jim McClure had arranged to discuss a proposal for the expansion. The meeting at the Mountain Home High School included ranchers, BLM officials, concerned citizens, Air Force representatives, and members of the Idaho congressional delegation. Along with other details of the proposal, Morris learned that an Air Force contractor had completed expansion plans without on-site inspections or searches of official BLM records to determine other uses and resources. A map of the proposal, though it showed no major landmarks, included a number of scattered remote radar sites. Morris took a copy of the map home and, comparing it to his own maps, figured out that the proposed range was in the Bruneau-Jarbidge River area south of the existing Saylor Creek range, an area with which he already was quite familiar. Since arriving in Mountain Home in the mid-1970s, he had spent much time hiking and camping in that area. He knew what was out there.

Unsure of how to organize opposition to the proposal, Morris spent the better part of July and August calling everyone he could think of. "How do you fight the Air Force?" he mused. Still, he played a key role that summer. Idaho Cattle Association head Gary Glenn asked Morris, who had been a legislative candidate, to arrange a meeting between Andrus and the cattlemen. Because of Governor Andrus' dislike of Glenn, Glenn could not do it himself. But neither could Morris. Instead, he arranged a meeting between the cattlemen and environmentalists. In a July 31 letter, he emphasized the importance of these two groups working together and urged them to attend the meeting set for the evening of August 14, 1989, in the public meeting room in the basement of the Idaho Supreme Court building in Boise.[11]

The road to Randall Brewer's Devil Creek Ranch, one of a cluster of ranches at Three Creek in the southeast corner of Owyhee County, runs right through the 1.5 million acres the Air Force wanted for the expansion of the Saylor Creek Bombing Range. Brewer was one of the sixty-four stockmen who grazed 14,500 cows and 11,000 sheep in the area. As a member of the National Cattlemen's Association Public Lands Council, Brewer eventually would take their concerns to

Congress to testify against the range proposal at a hearing before the House Natural Resources subcommittee on public lands. He knew the expansion would hurt his own operation and those of his fellow ranchers who leased most of their grazing land from the BLM.

Brewer received Morris' invitation. On the evening of August 14, he met with leaders from seven Idaho environmental groups. They agreed to put aside their differences and face the common threat. Brewer noted the irony of sitting in the same room with environmentalists. Morris compared the meeting to a family fight interrupted by an outsider. But they shook hands on it, and their handshake formed the foundation for what became known as "Idaho Is Too Great to Bomb," a group that eventually included American Indians, hunters, private pilots, and many others. The alliance also gave environmentalists some credibility. Idaho politicians were historically more likely to listen to a group that included ranchers. Folks at the meeting decided to organize a series of public information workshops in towns across southern Idaho in preparation for the anticipated Air Force public hearings.

The rest of southern Idaho heard about the proposed expansion from the state's congressional delegation. The Air Force said nothing. Nor did Governor Andrus, who in 1987 had formed a special task force to work with the Air Force on plans for expansion at the Mountain Home base and the Saylor Creek range.

Though details on Air Force plans had been sketchy that summer, in August a "Description of Proposed Action and Alternatives" was leaked. It was a planning document normally part of an environmental impact analysis process. Air Force officials, however, tried to disavow the document and would not release it to the public, saying a contractor had prepared it, not Air Force personnel. Bootlegged copies were photocopied and distributed by request from congressional offices.

The document outlined the proposal in detail. It showed a tactical range that could be arranged to represent anticipated battlefields of Europe, Russia, or the Middle East. The range would simulate actual conditions with front lines and defensive installations, such as anti-aircraft missiles. Next would be strategic and support targets, such as railroads, airports, and factories. The anti-aircraft defense would be simulated electronically by thirty or more portable field radar units strewn across the desert that would create an electronic combat range and score aerial dogfights. The strategic target areas, totaling approximately 250,000 acres, would be broken into four tactical ranges, with restricted access during operations. Approximately 64,000 acres would be used for live bombs and missiles and would be off-limits to people and livestock. The rest of the range would be open to recreational activities and grazing. Over all of it, pilots would practice evasive maneuvers that included dropping burning magnesium flares and chaff—bundles of aluminized silica fibers—to "fool" enemy radar and missiles.

A fighter-bomber's best defense is speed. And the planning document also called for designating all or part of the 7,000 square miles of military airspace over southwestern Idaho as a Supersonic Operations Area from the ground to 10,000 feet above the ground. Air Force policy does not allow supersonic operations below 30,000 feet except in specially designated areas—hence the large size of the proposal. Later versions of the range plans raised the airspace floor to 100 feet, with supersonic flight no lower than 5,000 feet—except in restricted airspace over target areas where supersonic operations would be from 100 feet up to 30,000 feet. At the lowest level, even subsonic fighter jets would be plenty loud—from 113 to 124 decibels—enough to cause physical damage to unprotected ears after a few minutes exposure. All the planes that would use the range were capable of supersonic flight. Supersonic operations were expected to result in up to forty sonic booms per day.

Like Meyr, however, most who heard rumors of the expansion plans expected something akin to doubling the size of Saylor Creek to somewhere around 200,000 acres. Nobody was ready for 1.5 million acres, live bombs, or low-level supersonic operations.

Chapter 3 / Sensitive Receptors

O N AUGUST 18, 1989, following a two-day tour of the proposed range, Idaho Sen. Jim McClure and BLM Director Cy Jamison asked Air Force officials to reduce their appetite for land, meet with the BLM, and compromise to avoid conflicts with ranchers as well as wildlife and recreation users.

No meetings were held. And that did not escape McClure's notice. The proposal would require congressional approval and would have to pass through McClure and his colleagues on the Senate Energy and Natural Resources Committee. Because the Air Force wanted to fly supersonic down to one hundred feet above the ground, the agency was required to own, lease, or otherwise control the land beneath such specially designated airspace. Only Congress could give the Air Force that control. Congress would decide whether there would be an expansion, what form it would take, and what restrictions and requirements it might include, McClure said.

The Air Force did not change its proposal. Instead, on August 22, officials announced that the first round of preliminary public hearings for an environmental impact statement would begin September 5. They hoped to issue the first draft of the impact statement in October. But that did not leave much time to conduct a thorough environmental analysis of such a large area with many users and complex issues. The rush, in fact, left a lot of people feeling like the Air Force already had made up its mind, that officials were only trying to jump through the appropriate hoops to satisfy the National Environmental Policy Act. The public comment period would end October 26, and the draft environmental impact statement would be out sometime in November, Cooper said.

"There's no way that you folks can adequately prepare an EIS, draft or any other kind, in the time frame that we're talking about," Glenns Ferry rancher Lee Presley said. "You can't even gather all the information that you need to write an EIS in that length of time."[1]

The Air Force was flying straight into a barrage of public opposition. That opposition began to coalesce with a series of public workshops in southern Idaho designed to inform people about the environmental impact process and how to prepare testimony. The real battle for the Air Force would be public relations, but that seemed to escape officials, including the officer leading the effort. The Air Force proposal was taking devastating fire in the form of Idaho Governor Cecil Andrus' flagging support, increasingly negative public opinion, and the dug-in opposition of cattlemen. Additional information released during the hearings and admissions by officials further weakened the Air Force's position.

Air Force officials tried to decoy the opposition by separating the plans to bring in F-4s and expanding the range from consideration of the real, on-the-ground effects of those changes and by hollow assurances that the two-part environmental impact statement would include everything people needed to know.

At the workshop in Glenns Ferry at the end of August, Herb Meyr met Brian Goller. Goller immediately recognized the value of Meyr's experience and special knowledge. Meyr could say when things made no sense, and he could identify false information. Goller would later describe him as "a pit bull who wasn't afraid to tell Air Force officials they were lying to the public."[2]

Brian Goller, in his late 40s, was no stranger to politics. His father was chief of staff for Sen. McClure and served on the Northwest Power Planning Council. The junior Goller had graduated with an education degree from the University of Idaho in 1977. He taught school for several years in northern Idaho and served on the Wallace City Council before returning to Boise to run the family publishing business. The bombing range issue turned him into an environmentalist. Following the August 14 meeting, Goller had helped organize and conduct the three grassroots workshops in an effort to provide people with information about the range proposal and encourage them to testify at the Air Force's upcoming public hearings. He wanted to make people realize the range proposal was a serious issue. And the number of those who showed an interest surprised him and lent a sense of excitement, a sense of the issue's importance.

Helping Goller was Kerry Cooke, a Boise activist involved with the anti-nuclear weapons organization Military Production Network. Through her activities, she met Grace Bukowski, Military Program director of Citizen Alert in Nevada, a group that had long been active in massive military land and airspace withdrawals in Nevada. Cooke invited Bukowski to Idaho to speak at the workshops.

Bukowski told of her experience with Navy expansion efforts at Fallon Naval Air Station in Nevada. She answered questions and explained the environmental impact statement process. And she gave people a sense of the bigger picture—that this kind of thing was happening in other areas. "Why is it OK to do to rural folks what's not OK for city folks?" she asked. Her answer was that it was not OK, and somebody needed to be told. The military needed training, but it also needed to be fair to those affected by that training.[3] By the time she was finished, Goller did not need to say anything.

After the Glenns Ferry meeting, Meyr went to the Mountain Home Air Force Base to talk to Lt. Col. Cooper—whom Meyr had instructed as a student navigator. Cooper, who was in charge of conducting the environmental impact statement for the changes at Mountain Home, and Mountain Home airspace manager Kent Apple briefed Meyr, showing him maps and details of the proposal. Meyr took these to the Idaho Department of Fish and Game in Boise.

Next he and a Fish and Game official visited the Boise District Bureau of Land Management office, then attended a meeting with Andy Brunelle of the governor's office.

Brunelle later recalled that the governor's office received its first sketchy details from the Air Force when Cooper showed them an autoclub roadmap of the large roadless expanse of eastern Owyhee County—but not the range. Following his meeting with Meyr and Fish and Game, BLM Boise District Manager Dave Brunner had gone to the governor's office to find out what was happening, Brunelle said. The following morning, a Fish and Game official showed up at the governor's office with a poster map and Mylar overlays that showed property ownership, resources, range developments, past fires. He completed the map with an overlay that showed the proposed range with its bomb impact areas and supersonic operating airspace.

When Governor Andrus learned that the proposal included live bombs and supersonic flight down to one hundred feet above the ground, he dropped his unqualified support for the proposal. Air Force brass had left him with the impression that supersonic flights would not be part of the package. Low-level supersonic missions were not compatible with existing uses in this area, he said. The confusion over such operations was understandable, however. At various times during 1989, Lt. Col. Cooper had said that the Air Force had no plans for low-level supersonic flight; that plans included supersonic flight from one hundred feet above the ground to ten thousand feet; and that there would be no supersonic flight below five thousand feet. Capt. Steve Solmonson—a public affairs officer at Mountain Home—claimed, incorrectly, that the F-4 was incapable of low-level supersonic flight.

People in southern Idaho were skeptical. All this, just for ninety-four aging F-4 fighters? Like Meyr, most people thought the Air Force was asking for more than it needed. Air Force officials responded that Mountain Home had to be ready to train the F-4s when they arrived from California, otherwise there would be no point in bringing them. The expansion was required to bring the planes, Air Force Capt. Wilfred Cassidy said.

But the Air Force had a lot more up its long blue sleeve. The F-4s would not be alone on the range. The planning document leaked in August revealed that every unit in all branches of the military would be invited to train at the new range. And the range was to be designed to meet the needs of future aircraft. The document envisioned bringing F-15 and F-16 fighters, B-52 and B-1B bombers, stealth bombers, and any new planes that came along to the new Saylor Creek range. Activity at the range would more than double. But when Cooper was asked about this future activity, he said he did not have all the details.

"That'll all be in the EIS."[4]

Lt. Col. James Cooper was the man in charge of the hearings and the environmental analysis of the range expansion. Cooper had earned a Distinguished

Flying Cross in the Vietnam War, where he had served as navigator in an AC-119G Shadow gunship. He had flown 185 combat missions and had logged seven hundred hours as a navigator. In the late 1980s Cooper was named director of realignment at Mountain Home Air Force Base.

When Cooper was born in New Rochelle, New York, on December 15, 1943, the Mountain Home Army Air Field was only a few months old. The Army had just acquired 420,000 acres south of the Snake River to establish the Saylor Creek Bombing Range. Cooper earned his wings in 1969 in navigator training at Mather Air Force Base in California and was sent to the 17th Special Operations Squadron in South Vietnam. By then the Air Force had long since renamed the Mountain Home Air Force Base and given back 300,000 acres of the Saylor Creek range. During the 1950s, the government built housing, warehouses, barracks, utilities, and runways. The base was home then to long-range B-29 bombers on Cold War alert and later, in the early 1960s, B-47s and three Titan missiles. By the time Cooper arrived, the base housed F-111s and electronic jamming EF-111s.

One thing apparently escaped Cooper. "He would look at us incredulously when we said we didn't think the Air Force needed more land," Kerry Cooke of the Snake River Alliance explained. "Col. Cooper personifies the Air Force mentality that if they just told the simple folk, we would see the wisdom of their plan. But the opposite happened. The more they told us, the more the plan didn't make sense. It looked more like just greed. The 'sensitive receptors' objected."[5]

More than 400 people showed up when the Air Force brought its proposal to Twin Falls for a public hearing in September 1989. Officials had set up a lecture hall at the College of Southern Idaho—a room that could hold approximately 125. But they quickly saw that they had misjudged the interest and moved to an auditorium across the campus. Turnout at the hearing was in large part the result of the workshops Goller had organized.

Once the hearing was under way, folks blasted the Air Force proposal. Janet OCrowley of Picabo walked up to the microphone and carefully turned it around to face the audience before she began to speak. But before she could launch into her comments, Cooper stopped her.[6]

"Could you please address the panel, please," Cooper said.

"OK," she replied.

"Thank you."

"I believe you can hear as well as anyone," OCrowley said, still facing the crowd.

"I would like you to face the panel please," Cooper said.

He castigated the crowd's outburst of laughter and asked everyone to cooperate with the panel and conduct the hearing in a proper manner. But people in the crowd yelled that Cooper should leave her alone and let her speak however she chose.

"I will not turn my back on the audience, and I will not turn my back on you, and I will direct my voice to both of you equally. Is that OK?"

Cooper grudgingly acquiesced.

At the next Air Force hearing, in Glenns Ferry, Randy Morris pulled a large, rusty metal object out of a paper bag—a dummy bomb that had missed its intended target at the Saylor Creek Range. He was not certain whether it was dangerous. But only moments after an Air Force officer said that the practice bombs it planned to drop on Owyhee County were no threat to public safety, another officer asked Morris to hand over the dummy bomb, saying it might be dangerous.

Morris had found the bomb in a popular campsite along the Bruneau River at the bottom of the thousand-foot deep Bruneau Canyon several miles outside the Saylor Creek range. Apparently it had missed its intended target by miles. The bomb, however, was not dangerous. It once contained a marker charge— similar to that in a shotgun shell—but that had long since been discharged. Morris did not know how long it had been there, but it raised concerns about a proposal that called for live bombs and missiles fired from five or six miles away, he said.

"I think it is absolutely insane to consider doing this, this close to populated areas and this close to areas that have been proposed by conservationists for national parks—an area that is widely used for hunting, fishing, and other recreational activities," Morris said.[7]

The area was also used by ranchers who were determined not to allow the Air Force to take over some of the best year-round grazing land in southern Idaho.

Dust boiled thick from under the pickup truck and hung in a rising plume that drifted eastward in the light breeze. Gravel banged on the underside of the truck as Randall Brewer rattled south toward Three Creek in the southeast corner of Owyhee County. The road, passable most of the time, cut through what would be the proposed Saylor Creek Bombing Range expansion. The pavement ended just outside the town of Bruneau, and Brewer was looking at forty miles of gravel road.

The land out there was laced with water lines and dotted with wells, storage tanks, and livestock water troughs, providing some of the best year-round grazing in southern Idaho. The ranchers and the BLM had invested millions of dollars in fences and water systems to water those pastures. Following expansive fires in years past, the BLM had also spent millions to re-seed vast areas with crested wheatgrass and other forage species, which had doubled the amount of livestock forage available in some parts of eastern Owyhee County. Here in the lower pastures, where the snow melted early in the year, Brewer's cows gave birth and fed on the early grasses. Other ranchers relied on the area for winter grazing. As the pastures greened up, Brewer would move his cattle up into the higher country of the ten-thousand-foot Jarbidge Mountains of northern Nevada. But like his fellow ranchers, Brewer was worried he would lose these pastures to the range expansion.

Air Force maps placed live-bomb areas directly over key grazing pastures, wells, and other grazing developments. One of the live-bomb areas was the spring calving pasture for 850 to 1,000 head of cattle. It would be like cutting one leg off a table, Three Creek rancher Bill Swan said. The leg by itself is not worth much, and the table is worthless without the leg. Miles of pipeline would be off-limits. Some grazing allotments would be cut into unconnected, inaccessible pieces. Though ranchers would be restricted from certain areas, they would be allowed onto most of the land, except during certain operations. But animals and water systems require daily attention, they said.

To minimize the effect on ranchers, the Air Force said it would try to locate the bombing areas away from major water systems. "Naturally you don't want us to put a five-hundred-pound bomb on top of your well," Cooper said. "We don't want to do that. It's a dumb idea."

Cattlemen thought the whole range proposal was a dumb idea. They suggested that Air Force officials admit their mistake in not assessing the resources within the area and start over somewhere else. J.R. Simplot Livestock Company foreman Tom Basabe suggested the Air Force look at Dickshooter Ridge on the Owyhee Plateau to the west.

"Any expansion which isolates this area or restricts use of this area would not be in the best interests for the livestock operators or for those concerned with the use and protection of natural resource values," Basabe told the Air Force.[8]

It was simple economics. The high desert summer grazing land such as Dickshooter Ridge could be replaced more easily than the productive year-round grazing land threatened by the Saylor Creek expansion. Without winter grazing, ranchers would have to feed more expensive hay to their cattle. At the time, grazing on federal land cost less than two dollars per cow and calf per month. But a cow eats a three-dollar bale of hay in about three days, more if it is cold. That adds up to about thirty dollars a month or more per cow.

In an effort to accommodate the ranchers, a "shared use" plan was developed. The Air Force would use the range from 7 a.m. to 10 p.m. daily and about half the weekend days per year. Ranchers and others would be free to use the area the rest of the time. It was the Air Force's idea of multiple-use, Cooper said. One suggestion was for ranchers to check their cattle at night. Ranchers ridiculed the proposal. Some thought it was a joke.

The Air Force also proposed to buy the fifteen thousand acres of private land within the expansion boundaries. But even if they were compensated for land lost to the range expansion, many ranchers complained they would not be able to replace key pastures. In most cases, only grazing permits would be lost and compensated. But if lost grazing meant the ranch would not be able to operate, the government would offer to buy the whole ranch, said Aden E. Hamilton, assistant chief of the U.S. Army Corps of Engineers real estate division, assigned to assess the value of ranch losses for compensation or purchase.

The ranchers did not oppose the country's defense needs, but like many others in southern Idaho, this was land they cared about. And they were loath to see their livelihoods disappear. They feared the new range would drive them from homes and ranches that, in some cases, had been in the family more than a hundred years. They would not give up without a fight.

Before the first round of public hearings had concluded, Air Force officials recognized they were in trouble. And in a move to deflect some opposition, they announced that the upcoming environmental impact statement would be split into two parts. A draft of the first tier would be released in December and public hearings conducted in January 1990. Tier One would evaluate the effects of bringing the planes to Mountain Home and the potential for expanding range capability in Idaho. Tier Two would consider the detailed effects of such a range expansion. Once the second tier was done and the Air Force had issued a Record of Decision, the Air Force would go with the BLM to Congress to get withdrawal of the public land.

The Air Force—which had tied the expansion to the arrival of the ninety-four F-4s—now admitted that the planes would come whether the range was expanded or not. The expansion was "not necessary to support the F-4 mission," said Deputy Assistant Secretary Gary Vest, admitting the Air Force had done a poor job of informing the public about its expansion proposal before public hearings.[9]

Despite Vest's reassurances, however, people remained skeptical of the two-part impact statement. Considering the benefits of expansion in the first stage and the costs in the second skewed the decision-making process in favor of expansion. The hearings became polarized and emotional. Critics derided them as nothing more than a waste of time, because the decision already had been made. Some people charged that in their rush to hold perfunctory public meetings on the expansion, Air Force officials had neglected to properly inform the public. One man complained that the Air Force had not provided answers on the effects of low-level supersonic flight.

"You have that information, you should have given us that before this meeting," Glenns Ferry resident Robert Hall told a panel of Air Force officials at the crowded local high school gym.[10]

Cooper replied that the Air Force "can't assess that until we know the design of the range and target arrangements."

Hall charged that Air Force officials knew well the effects of a fighter jet flying above the speed of sound at five hundred feet above the ground. That kind of information would be needed if people were to make intelligent comments in the scoping process, he said.

But low-level supersonic flight would not be covered by Tier One, Cooper replied. It would cover only impacts of supersonic flight from five thousand to thirty-five thousand feet above the ground. It looked at the big picture, at the effects of increased air traffic and the requirements for a training range. Tier Two would look at specific details of a training range, including low-level supersonic flight between one hundred feet and five thousand feet above the ground, target placement, and range boundaries. The two-tiered approach would allow more detailed study of the expansion effects and more opportunities for public comment, Cooper said.

People were not convinced.

The range expansion faced other problems. The BLM had proposed wilderness designation for 20,800 acres of the Bruneau River-Sheep Creek Wilderness Study Area and 16,740 acres of the Jarbidge River Wilderness Study Area—all of which would be included within the expanded range. And the population growth in southern Idaho, and particularly in Boise, meant more people would be looking for wide-open spaces for recreation.

Idaho Department of Fish and Game officials asked Air Force planners to delete from the proposal the proposed airspace expansion over the canyons of Big and Little Jacks creeks. This area was potential wilderness, habitat for an important bighorn sheep herd, and critical winter range for mule deer, pronghorn antelope, and sage grouse. Because the area had been outside the existing military airspace, it was valuable to researchers as an area against which to assess the effects of military overflights on wildlife in other parts of southern Owyhee County.[11]

Cooper acknowledged that Air Force officials knew about the recreational use in the Bruneau and Jarbidge river canyons and the other rivers, and they knew about the wildlife and the amount of grazing there. But if they knew all that, a number of people asked, why did they make this proposal right in the middle of what it now acknowledged as prime recreational and productive grazing land? A bombing range was not appropriate for an area so developed and heavily used, they complained. In response to the public outcry, the Air Force decided to slow down, to let people digest what was going on.

In early January 1990, lines on the map of the proposed expansion were symbolically erased. Cooper said the Air Force would study all of Owyhee County for suitable sites to locate a new bombing range.

But no one was fooled.

To diffuse additional opposition to its proposal, the Air Force also scrapped its plans for low-level supersonic flight. It would limit such flights to five thousand feet above the ground. The expansion would be the culmination of the expansion proposal begun in 1984, and it would accommodate unspecified future Air Force needs, Cooper said. But exactly how much land and airspace the Air Force "needed," or exactly what it was the Air Force wanted in Owyhee County, he would not say.

"We know exactly what we need. It'll all be in the EIS."

Chapter 4 / Opposing Forces

To TRY TO CALM an increasingly critical public, and to give people the feeling that they were to be part of a compromise, U.S. Rep. Larry Craig of Idaho had convinced the Air Force and the Bureau of Land Management to convene a "working group" of Idaho citizens. The group was supposed to comment on the Air Force proposal, define issues and questions to be answered in the impact statement, and suggest alternatives. It would submit its conclusions and recommendations to the Air Force as part of the public comments after the draft environmental impact statement was released in early 1990.

But it became immediately obvious at the opening meeting in the fall of 1989 that the group was expected to reach a compromise favorable to the Air Force. U.S. Sen. Steve Symms had told the participants, "If you're at this meeting for *no* bombing range, you're at the wrong meeting."

Critics suggested that the group was created so Craig would look good politically and have some influence over the final decision. But it might be that Craig knew a lot more than he was letting on, such as the push to establish the range in a timely manner had less to do with "combat readiness" than it did with using the realignment of the F-4s as a lever to expand the Mountain Home base before the planes were retired from the Air Force inventory.

The group was to work out possible alternatives to the expansion, which was to be examined in Tier Two. The group would continue meeting and discuss the Tier One draft when it was issued. But to be effective, the group needed to know specifics of the Air Force's plans and needs, and those would be in Tier One, which was not available when the group started meeting. Air Force officials promised to make all requested material available to the working group. They did not.

To facilitate the group's work, officials should identify what planes using what weapons would use the Saylor Creek range, said retired Lt. Gen. Lynwood Clark of the governor's expansion task force. They did not.

Working group members complained that the Air Force did not provide the information they requested. This lack of cooperation made informed participation more difficult and more time-consuming—if not impossible. But providing information was not in the Air Force's perceived self-interest; the working group would have forced all the cards on the table. The group also commented that the Air Force had failed to adequately justify the need for the proposed range expansion. The group's nearly unanimous conclusion was a suggestion to drop the proposed expansion and halt the impact statement process.

Several participants said that the confusion over the range proposal originated with the Air Force, which had continued to treat the range expansion and the transfer of F-4s to Idaho as one process, using the transfer to "leverage the expansion." Others characterized organizing the group as the Air Force's effort to manipulate a dissenting public into developing a compromise. In that it failed miserably. The working group concluded that the expansion was not justified. With the information available, participants could not reasonably analyze what the Air Force needed and therefore could not form a compromise. Under the circumstances, they made the only recommendation they could. Craig scolded the participants for failing to reach a compromise, and Mountain Home representatives accused the group of being biased and failing in its charter. The *Mountain Home News* said, "This group failed miserably in its charge and in doing so seriously threatened the long term existence of this base and this community."[1]

The only working group member to stand by the Air Force was Mountain Home real estate agent Jack Streeter, who also castigated the group, saying that "nobody moved an inch from their original positions."[2]

But it was the Air Force, not the BLM, that scuttled the working group, even though the BLM later was labeled obstructionist by the Mountain Home Chamber of Commerce. Had the group been allowed to continue its work, the Air Force might have gotten its bombing range, one BLM official said.

The group worked through the winter and submitted its final recommendations as part of the environmental review process during the public comment period. But Air Force officials would not comment on them until the comment period closed.

"It would be inappropriate to comment during the comment period," Lt. Col. Cooper said.

Also working over the winter of 1989-1990 was Betsy Buffington of The Wilderness Society's Boise office. Following the public scoping meetings the previous fall, she had started writing to others who opposed the bombing range expansion, asking them to join a small group already meeting once a week at the society's office. Those who met represented various interests, loosely unified in their opposition to the range. But by working together, they could pool resources and information. Most of those who came to the earliest meetings already were involved with environmental issues. Buffington herself had ties to the desert in Owyhee County. In the early 1980s, her father, former state BLM director Robert Buffington, was removed from his position by Interior Secretary James Watt for proposing wilderness in Owyhee County.

Reluctant at first, Herb Meyr became a regular member at those meetings. The high desert of Owyhee County was a great place to fly, he said. But it was the wrong place to put a bombing range—it was not a dry lakebed like many other deserts in the West, nothing but rocks and sand. Supersonic operations should not be done over land used for recreation or livestock grazing. To Meyr, the Air Force's proposal would be like building a bombing range along the Grand Canyon with bomb impact areas three miles from the rim, dropping live ordnance, then saying it would not affect people, recreation, or wildlife.

Also joining the weekly meetings were ranchers and members of the Foundation for North American Wild Sheep, the Aircraft Owners and Pilots Association, and the Native Plant Society, among others—as many as fifty groups and individuals eventually became part of the coalition. Until the Saylor Creek issue, people who worked on desert issues had mostly worked alone. They were not alone anymore.

A similar group began to meet in Twin Falls, east of the proposed bombing range. Janet OCrowley coined the name "Idaho Is Too Great to Bomb" for the loose-knit opposition. She was a tireless and outspoken opponent who had worked with Randy Morris from the beginning. Both groups organized and solicited participation in the effort to oppose the Air Force proposal, and they helped people prepare testimony for public hearings. OCrowley urged all opponents to speak a simple message with one voice against range and airspace expansion, the use of live ordnance, and supersonic flight. One important topic that came up often at both groups was how to reach out beyond the small core group of individuals.

Bob Stevens of Ketchum did just that. Stevens had no inhibitions about approaching people—and he had his airplane. He would get in it, fly somewhere, and meet with people face to face rather than talk to them over the phone. He gathered information and got it to the right people, and he got those people to listen to him. In his business of dealing in real estate in the Sun Valley resort area of central Idaho, Stevens had learned to move in a variety of circles that included the wealthy and influential. He put that talent and his connections to work in the fight against the Air Force. He was not often visible, but he was a key player behind the scenes.

In his early fifties, a fringe of graying, curly hair around his bald pate, Stevens' easy manner exuded competence. He had learned to fly at age seventeen in Pottstown, Pennsylvania, in a small, light biplane—the kind where someone had to hand-crank the prop to start the engine. He eventually became a naval aviator and, after the Navy, a commercial pilot. But he quickly bored with civilian flying and turned to real estate instead.

"I didn't want to be a bus driver for the rest of my life," he said.

Peering through the rain-streaked windshield of his Cessna 185, Stevens said he opposed the bombing range expansion because "a wonderful area was going to be ruined by something that didn't need to happen." An avid bird hunter, Stevens had spent many hours in Owyhee County with his Brittany spaniels, hunting chukar. And as a private pilot, he was aware of the effects the huge aerial battleground would have on the ten thousand or so civilian flights that cross southern Owyhee County every year. Though for four years Stevens all but dropped his family and his business for his fight against the range, he said, "It's one of the most important things I've done in my life."[3]

The weekly meetings helped unify and coordinate public opposition to the range. With the efforts of such groups and the work of key individuals, this opposition continued to grow during the first half of 1990.

The Air Force released Tier One of the environmental impact statement at the end of February 1990. But it did not say exactly what the Air Force needed, as Lt. Col. James Cooper had promised. When asked about this, Cooper replied that many effects associated with details of a range expansion would be covered in Tier Two. Tier One recommended expanding the 109,000-acre Saylor Creek range and suggested that an optimum range would include 7,500 square miles of airspace and nearly 2 million acres of training ranges to provide realistic pilot combat training. Increasing demands of units training at the range and additional planes coming to Idaho had outpaced the range's current capacity, Tier One said. For pilots to train properly, the range should include supersonic flight above five thousand feet and a place to drop real bombs. The range would have to be flexible to prepare for what Gary Vest had called the long-term challenges of the future and to meet the training requirements of the planes that would replace the F-4s.

The release of the impact statement opened the public comment period, and public hearings were set for early April. But before the public hearings began, a number of civic leaders and others asked the Air Force for a demonstration of a sonic boom.

Approximately seventy-five people gathered at the Saylor Creek Bombing Range early one morning in March 1990. The sudden crack, though anticipated, startled the group. As much felt as heard, it was little louder than a shotgun going off nearby, but it came from an F-4 flying straight and level at 790 mph, just over the sound barrier, about five thousand feet above the ground. Still, it formed no comparison to the sonic booms from fighter jets making evasive maneuvers over the backcountry, and such booms would be a regular thing over the range—an average of more than forty a day of varying intensity.

Sonic booms can have a profound effect on people's lives. On March 3, 1983, a Navy F-14 Tomcat from Fallon Naval Air Station, Nevada, maneuvering about six thousand feet above the ground caused a sonic boom that shook loose the corrugated steel roof on Ed Robbins' home in Dixie Valley, Nevada. Robbins, a former civilian aviation electronics and armament technician with the Air Force, said he eventually was driven from his home, located under a supersonic training area of fifty-five hundred square miles in northwestern Nevada. Supersonic operations above Dixie Valley had regularly cracked windows and walls. Acknowledging the health effects associated with supersonic operation, including nervous disorders and depression, the Navy bought many of the homes in the valley. When the residents had gone by 1987, the Navy bulldozed and burned the houses. Robbins moved to Sagle, Idaho, and he blames his wife's chronic depression on their exposure to sonic booms.

A study by the state of Nevada concluded that no well-accepted evidence showed that sonic booms cause nonauditory health effects on humans. A U.S. Navy study, however, reported that repeated sonic booms can lead to health problems, including nervous disorders and depression. It also said that booms can severely startle people, causing them to lose control of vehicles and power equipment. Sonic booms from low-flying aircraft are likely to result in property damage and a reduction in property values.

As an aircraft moves through the air it produces pressure waves similar to a ship's wake. At the speed of sound—about 750 mph at sea level—those waves are concentrated into a single shock wave that trails the plane in the shape of a cone. A sonic boom is the sudden change in pressure as that shock wave passes. The shock wave's intensity is expressed as "overpressure" in pounds per square foot— simply the increase above normal atmospheric pressure. The change of a few pounds per square foot is similar to the pressure change in an elevator as it descends two or three floors—but in a fraction of a second.

The intensity of a sonic boom varies with the plane's altitude and maneuvering and with the weather. For aircraft in level flight, sonic boom overpressure varies from less than a pound to about ten pounds per square foot. Peak booms for maneuvering jets are two to five times higher, but over a much smaller area. The strongest sonic boom ever recorded was 144 pounds per square foot produced by an F-4 flying just above the speed of sound at one hundred feet above the ground. Typical sonic booms from F-4s range from 2 to 6.8 pounds. Under realistic flight conditions, the strongest boom likely to occur is 21 pounds per square foot. That could cause some damage—shattered glass, for example. Buildings in good condition are unlikely to be damaged by overpressures of less than 16 pounds, according to Air Force information.[4]

In 1989, Air Force officials told Idaho residents that modern supersonic aircraft normally produce booms from less than 1 to approximately 5 pounds per square foot. Flights at more than five thousand feet above the ground would result in booms from .5 to 1 pound per square foot, and generally structural damage from booms begins at above 11 pounds per square foot, officials said. But a study done by the state of Nevada showed that minor structural damage begins between 2 and 5 pounds.[5]

Another study conducted by Nevada for the U.S. Navy showed that sonic booms between 1.5 and 2 pounds result in a significant public reaction. Exposure to eight to ten sonic booms per day of about 1.7 pounds per square foot were rated as unacceptable by twenty-six percent of the residents near Edwards Air Force Base.

Hundreds of people spoke at the public hearings and wrote to the Air Force in the spring of 1990. Most of them opposed the use of live ammo, low-level supersonic flight, and the effects of the bombing range on recreation.

The Snake River Alliance was fresh from battle against the federal Department of Energy, having beaten back a proposed plutonium refinery in eastern Idaho, and it was looking for another fight. Who better than a peace group to question the need for a new bombing range where Air Force pilots train for war? So asked Liz Paul, the alliance's executive director. Outrage over the proposed plant that would produce nuclear weapons material drove her to become an activist. She applied much of what she had learned about fighting the Department of Energy to the bombing range issue. And living in Ketchum, she had learned much about that issue from Bob Stevens. Rarely in the limelight on the bombing range issue, Paul, like Stevens, was also effective working behind the scenes. When the Air Force refused to hold hearings in Paul's stomping grounds, the Wood River Valley, she persuaded local government officials to conduct their own public hearings. The comments were transcribed and submitted to the Air Force to be included in the environmental impact statement.

The Northwest Pipeline Corporation was concerned about its regular aerial patrols of the natural gas pipeline that ran down through Owyhee County near the proposed range. Federal Animal Damage Control officials expressed similar concern about responding to predator problems in a timely manner. Even the federal agency that managed the land, the normally pliant Bureau of Land Management, opposed the proposal. Agency officials criticized the two-tiered approach. They also noted that the Air Force did not explain its training needs to justify the proposal. The impact statement did not look at national training needs or the needs of other planes that would use the range.

"If they need it, the Air Force has to explain it better," said Dave Brunner, Boise District manager of the BLM. The proposal also included supersonic flight and airspace changes in the no-action alternative, but those changes in operations were training issues and should have been covered in Tier Two, he said.[6]

Idaho Congressman Richard Stallings greeted with mixed feelings the suggestion that the Air Force start over. No matter how much time Idahoans are given, he said, they still would balk at the Air Force locking up 1.5 million acres of public land and pounding it with live bombs.

But not everyone opposed the Air Force's proposal. The Mountain Home Chamber of Commerce and its Military Affairs Committee of Fifty fought for the proposal because, members said, the economy of Mountain Home would become more stable as a result of realignment and a range expansion. A new, larger range would greatly increase the likelihood that the Air Force would stay in Idaho long after the F-4s were retired. Members also claimed that environmentalists who opposed the range were not as interested in protecting the desert as they were in creating an elite playground for wealthy urban backpackers at the expense of local residents trying to make a living off the public land. But nothing could have been further from the truth. Among the staunchest opponents of the expansion proposal were Owyhee County cattlemen—some from the Mountain Home area.

The Air Force wrapped up public comment on the Tier One draft impact statement on May 15, 1990, and officials expected to release the final impact statement and record of decision in June. Public hearings on Tier Two would begin in late summer or early fall, depending on the decision reached in Tier One. The size of the expansion, training needs, effects, and alternatives would be covered in Tier Two.

But the proposal already had begun to unravel. Earlier that month, Mountain Home Wing Commander Col. Victor Andrews told the *Washington Post* that if the range were not expanded, training the F-4s would be difficult. The level of their combat readiness would be low, and eventually Mountain Home would not be very attractive as an Air Force base.[7] A few days later, however, the *Post* revealed a secret Air Force memorandum calling for the retirement of all its F-4s. Air Force officials refused to talk about the "Program Objectives Memorandum." Each military branch had sent such a memo to Defense Secretary Dick Cheney in May 1990 in response to Cheney's call for defense budget cuts between 1992 and 1997. The memorandum was the Pentagon's plan to meet these reduction targets.[8]

If it were true, Cooper said, the Air Force would have to modify the impact statement. Air Force Pentagon spokesman Capt. Sig Adams said the Program Objectives Memorandum was a classified Defense Department document, and he was not privy to such secrets. But the expansion of the Saylor Creek range might go ahead even if the F-4s were retired, he said. All Air Force programs are subject to review, and the Air Force would not stop the environmental review process based on a decision that had not yet been made.

Congressman Larry Craig pressed Cheney for clarification on Air Force intentions for the F-4s and the future of the Mountain Home base. Grudging support for a scaled-back range expansion turned to apprehension when it was learned that the F-4s might not be coming. Citing the importance of a good relationship with the Air Force to the economic stability of Mountain Home, Craig said he would push Cheney and Rice to identify future missions for Mountain Home—missions that would last into the twenty-first century.

Cheney told the Idaho congressional delegation and the Mountain Home Military Affairs Committee that all the F-4s at George Air Force Base might not move to Mountain Home. And on May 18, one week after the public comment period on the impact statement had ended, Cheney extended it to September 15 to "ensure the best possible decisions are made."[9] On May 23, Air Force Secretary Donald Rice told the Idaho delegation that the F-4s would probably be retired soon, and it would not make sense to move them just to be retired at a different base. Options for the Mountain Home base included moving all the Air Force's electronic combat planes there to closing the base. Cheney then extended the public comment period to January 1991, and the Defense Department announced a moratorium on further military land acquisitions. "The department must ensure that we propose the acquisition of land only where there is a clearly demonstrated need."[10]

The Saylor Creek expansion appeared dead.

But at a meeting in May in a motel conference room in Mountain Home, Herb Meyr said Gary Vest had told him public opinion would not stop the Air Force. "We might not need it now. But we might in thirty years, and we won't be able to get it then because people will be down there—more and more people using it, and they won't want to let it go."[11]

Part Two:

Idaho's
Proposal

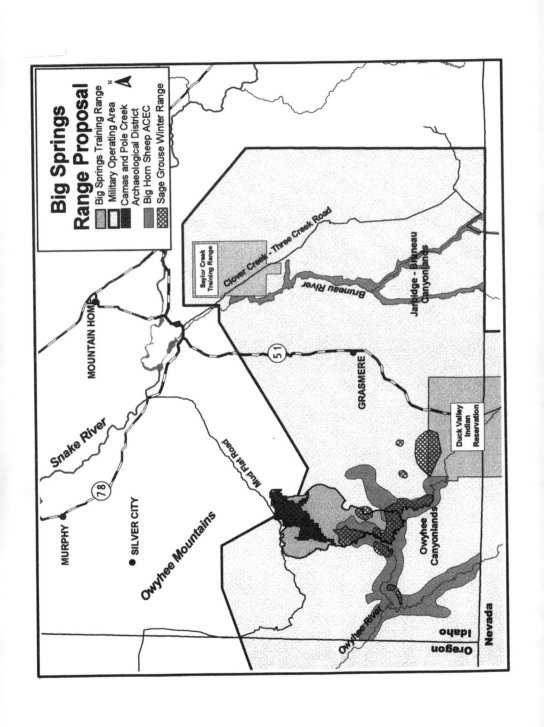

**Big Springs
Range Proposal**

N

- Big Springs Training Range
- Military Operating Area
- Camas and Pole Creek Archaeological District
- Big Horn Sheep ACEC
- Sage Grouse Winter Range

MOUNTAIN HOME

Snake River

MURPHY

78

SILVER CITY

Owyhee Mountains

Mud Flat Road

Saylor Creek Training Range

Clover Creek - Three Creek Road

Bruneau River

Jarbidge - Bruneau Canyonlands

51

GRASMERE

Duck Valley Indian Reservation

Owyhee Canyonlands

Owyhee River

Oregon
Idaho

Nevada

Chapter 5 / A Salable Thing

T HE DEMISE OF EXPANSION PLANS at the Saylor Creek Bombing Range in 1990 sent Idaho scrambling. The F-4s were not coming. The EF-111s stationed at the base were leaving. And the point man on the governor's military expansion task force, retired Lt. Gen. Lynwood Clark, had warned that a modern viable range was the only reason to keep Mountain Home Air Force Base open. Residents in the town were scared.

But with the help of Gov. Cecil Andrus, Gary Vest, and Dave Jett, a former motorcycle shop owner from Mountain Home, the effort to expand training facilities in Idaho merely went underground. Andrus had taken to heart the implied threat in Col. Victor Andrews' comment that if Mountain Home did not have proper training facilities, the base would close. Thus the plan had been simmering in his office since mid-1990 to convince the Air Force why Idaho—with good flying weather, low cost of living, and plenty of room—was a good place to locate a base. "We think it's a salable thing," Andrus said. "In the meantime, airplanes are still leaving Mountain Home."

He saw that Idaho would have to compete in the years ahead with other states that were also trying to keep their bases open in the face of tightening defense budgets and new rounds of base closures. The state would have to show the Air Force what it had to offer, and part of that would be an improved bombing range at Saylor Creek—or somewhere, he said. He only hinted at his plan that would include electronic ranges and a combination of state-owned and private land. His intention was to make sure the Air Force had what it needed so it would stay.[1]

The Air Force probably had a genuine need for upgrading its Idaho training facilities and the bombing range that was designed for World War II heavy bombers. It was not appropriate for modern high-speed jet fighters. But the Air Force did not do a good job of researching the area it proposed for the failed bombing range expansion. It was heavily used and thus not an appropriate place for warplanes to practice. But by the time the proposal reached the public, any real need had been clouded by the realignment of the F-4s. The transparent effort to link the two only bred suspicion. And so it was with suspicion that Idahoans received the governor's proposal.

Word of Andrus' plan leaked out in early February 1991. Though the governor's office would not talk about it publicly, Air Force officials in Washington, D.C., already had given it a favorable review. Back in July 1990, the governor had told Defense Secretary Dick Cheney that representatives of ranching and environmental groups would be willing to work out a solution to the base's training needs. On September 24, Andrus had suggested that the state give the Air Force access to state lands "for use as tactical training weapons impact areas"—a bombing range. On December 20, Andrus had sent deputy assistant Air Force Secretary Gary Vest a detailed proposal for a training area that would provide "inert ordnance delivery and electronic training" close enough to Mountain Home to permit maximum training efficiency. The location was about fifty miles southwest of Mountain Home, north of the East Fork of the Owyhee River.

Vest had replied on January 11, 1991, that Andrus' proposal offered significant improvements in training flexibility over the Saylor Creek range. But he cautioned that expansion at Mountain Home still was on hold—subject to a moratorium on military land withdrawals. Andrus noted that his proposal relied on state land acquired by trade, not withdrawal of federal public lands. Still, the Air Force would have to buy the private land within the proposal. By January 31, the proposal had been completed and forwarded to the Secretary of the Air Force for consideration. Andrus had proposed that the state would provide a block of land large enough for a tactical bombing range. The state would own it; the Idaho Air National Guard would operate it; and the Air Force would lease it.

In his proposal to the Air Force, Andrus had promised the cooperation of Idaho environmental groups and the Idaho Cattle Association. But members of those groups said he had not consulted them. He had offered state resources without consulting the people of the state. Though he had been working on it for seven months, he did not make the proposal public until February 1991. And he had angered a powerful segment of the hunting community. The Foundation for North American Wild Sheep said the proposal disregarded thirty years of efforts to establish and maintain a herd of California bighorn sheep in the canyons of the Owyhee River's East Fork.

Cecil D. Andrus was popular in camp on hunting and fishing trips because he never minded cooking or doing the dishes, one friend said. Andrus, "Cece" to his close friends, was born on a ranch outside of Hood River, Oregon, on August 25, 1931. He graduated from a Eugene high school in 1948, the year he was married, then attended Oregon State University. During the Korean War, he served in a U.S. Navy patrol bomber; afterwards, with three hundred dollars and a 1949 Chevy, he moved to a job at a sawmill in Orofino, Idaho. He fell in love with the streams and mountains of northern Idaho, where he fished and hunted.

When Andrus was elected to the state Senate at age twenty-nine in 1960, he had never been to the state capital. He represented Clearwater County there for eight years. In 1970, he ran for governor on the Democratic ticket. At the time, he had just been promoted to state manager of the Paul Revere Life Insurance Company. He became the first governor in the country elected to office on an environmental issue. He had fought against plans for an open-pit molybdenum mine at the foot of one of the tallest peaks in central Idaho's White Cloud Mountains.

"You can't hunt, fish, or picnic in an open pit mine," he said.[2]

Working with U.S. Sen. Frank Church of Idaho to establish the Sawtooth National Recreational Area and the largest chunk of wilderness in the lower forty-eight states—the Frank Church–River of No Return Wilderness in the central Idaho mountains—got Andrus noticed at the Interior Department. He was reelected governor in 1974, but he resigned in 1977 to serve as Secretary of the Interior under President Jimmy Carter. Andrus made a name for himself with his work in the 1978 campaign to preserve the country's largest remaining wilderness. He traveled throughout Alaska, facing tough, hostile crowds. Unflinching, he convinced most Alaskans that the proposed wilderness would not lock them out of their own state.

He returned to Idaho in 1981 and in 1986 was once again elected governor. He attributed his success to knowing when to quit. With a fringe of silver hair rimming his head, tall and straight as a ponderosa, he cut a patrician figure. He was quick to laugh, quick to anger, and had a long, unforgiving memory. He also had a reputation as a consummate politician with a conservationist bent. During his fifteen years as the Democratic governor of a rock-ribbed Republican state, Andrus showed that a healthy environment did not have to come at the expense of economic development. During his tenure, the state's economy grew steadily, yet he fought for protection of more than one million acres of wilderness in Idaho, battled the Department of Energy over nuclear waste, and kept coal-fired power plants out of the state. Andrus recognized the importance of compromise, but he was not a good loser. Always willing to risk his political neck for what he thought was right, he was as unyielding as a mother grizzly.

But Andrus' stance on the bombing range was the one time when many thought he should have quit. He found himself on the other side of the familiar environmental dilemma of balancing economic stability and conservation, for without a new bombing range, which threatened a fragile high-desert ecosystem, the Air Force implied it would close the base that formed the mainstay of Mountain Home's economy.

When he had returned to Idaho in 1981, he had brought back many expectations based on his accomplishments as Secretary of Interior. Yet in the final days of his public career, Andrus fought a battle that brought rancor even from his friends. A lot of people who had supported him over the years felt he had let them down—hunters and fishermen, conservation groups, those who had backed his fight over radioactive waste and his efforts to save Idaho salmon, and those who voted for him in his narrow victory in the 1986 election.

Perhaps the most bitter disappointment of all was that despite his own assurances that the range would not harm wildlife or the environment, the proposal was opposed by the Idaho Department of Fish and Game and his best friend and hunting partner, Fish and Game Commissioner Richard Meiers.

But Andrus was receiving help from another quarter. Ever since the Air Force had started looking at ways of expanding its training facilities in southern Idaho in the early 1980s, Dave Jett had been involved. With financial help and advice from the Air Force to become less dependent on the military, the Mountain Home Chamber of Commerce had started an effort at economic development in Mountain Home. Jett, who had lived there since 1979 and ran a motorcycle franchise, was a member of the chamber steering committee.

As the threat of the base closing increased, the interest level in it went up, Jett said. He approached Andrus with an idea of how to maintain the base's attractiveness to the Air Force, convincing the governor that the state needed to become involved in keeping the base open and to use existing facilities as assets. Idaho was weak politically and must use what assets it had to keep the Air Force in Idaho, Jett told Andrus. The Air Force had said it wanted to expand its training facilities in Idaho. The state could be a proponent of that effort. Jett sold the motorcycle shop in 1990, when he went to work for the state Department of Commerce as the governor's liaison in support of the proposed range.

The seeds for the governor's proposal were planted long before the Air Force shelved the Saylor Creek expansion. Andrus had said then that the continued existence of the Mountain Home base would be ensured by an expansion. But the largest of the private landowners in the area, Idaho industrialist J.R. Simplot, was adamantly opposed to the Saylor Creek proposal, Simplot foreman Tom Basabe told Andrus. Simplot's company was one of the largest outfits that would have been affected—with more than twenty-five thousand animal unit months (a unit of grazing management that equals the amount of forage a cow eats in one month) in the Bruneau and Jarbidge resource areas. Simplot would oppose any proposal that affected the land no matter who backed it, Andrus said. But Basabe had some other ideas, and he knew some ranchers who would go along. He suggested looking at the 600,000 acres of the Big Springs-Dickshooter Ridge area to the west, an area with fewer private interests and grazing developments.

Moving the range to the Dickshooter Ridge was as much the idea of Joe Black and Sons as Simplot's, Chris Black said. The Saylor Creek proposal also threatened the heart of the Blacks' Bruneau-based operation. The family's operation on Dickshooter Ridge was less productive, the land there available only part of the year. The lower elevation operation that would have been affected by the Saylor Creek proposal made up about two-thirds of their operation and provided winter grazing and spring calving.[3] The Blacks owned about 3,700 acres on Dickshooter Ridge, and they leased about 95,000 acres of BLM grazing land. The private land controlled access to the public land on which the range would have been located. And the range would have cut the operation on the public land in half, making it impractical to operate. Though the ranch had been in the family since the 1940s, the Blacks decided it would be better to sell out and buy another ranch, and they offered their land to the state and the Air Force.[4]

Robert T. Nahas—a friend of Andrus and a wealthy entrepreneur with holdings throughout the West—also owned land on Dickshooter Ridge. Nahas claimed to be a friend of Air Force Secretary Donald Rice, with whom he had served eight years on the Board of Directors of Wells Fargo Bank. Nahas, too, told the governor that he would be willing to sell his 2,500-acre Deep Creek Ranch. Andrus liked the idea. Two willing sellers were much easier than having to go through the condemnation process, which could take ten years in court with no guarantee of winning. And he would not have to fight Simplot.

While war raged in the Persian Gulf in early 1991, Andrus wrote to key members of Congress seeking support for his range proposal on Dickshooter Ridge and for the Mountain Home base. On February 8, Andrus publicly released his range proposal, dubbed the Big Springs Training Range—only four days after Dave Jett had insisted the proposal was still only in the concept stage. It was scaled back from the 1989 Air Force proposal yet large enough to accommodate realistic tactical training, Andrus said. "Every important consideration for the protection of the high desert ecosystem, the values of Idaho citizens, and the health of the Idaho economy are addressed in this new plan. The truly sensitive canyonlands of the desert expanse have been carefully excluded from the training range, the use of live munitions is prohibited, and the Air Force still will have an essential training facility near Mountain Home Air Force Base."[5]

He made it sound like a benign proposal that would save Mountain Home, protect wildlife, and maintain public access to popular recreation resources. Andrus' assurances notwithstanding, the proposal was deceptive. Because low-level supersonic operations and live bombs were not included in the governor's proposal, the Air Force did not need to own or control the ground beneath the airspace. That was one reason the Air Force had asked for 1.5 million acres in the failed Saylor Creek expansion proposal. Without the need for a large buffer area, the proposal looked much smaller. But moving target areas to the west, away from the existing Saylor Creek range, spread the electronic combat range over three million acres of public land in southern Owyhee County—twice the size of the Saylor Creek expansion.

The proposal was not popular. But with war raging in the Middle East and with Andrus backing the proposal, things looked bleak for the opponents. How could they explain to the public that it was not really a smaller proposal? The feeling was that if Andrus liked it, it must be good. And it was hard to attack the need for training during the Gulf War. People were reluctant to stand up and have their patriotism questioned. The proposal was in most respects identical to the Saylor Creek range expansion proposal, with the exception of low-level supersonic flights and live ordnance—and it, too, was about to become tangled up with a realignment proposal by the Air Force.

On April 12, 1991, less than a year after the demise of the Saylor Creek expansion, the Air Force presented to the Base Closure and Realignment Commission a proposal to establish a new wing at Mountain Home. Because of the success of its "composite wing" in the Persian Gulf, the Air Force was considering the concept for a number of U.S. bases, including Mountain Home.

Unlike a conventional wing, made up of a single type of airplane, the composite wing would be made up of a variety of aircraft that perform the various functions required to fight a battle—strike aircraft to attack ground-based anti-aircraft defenses, fighter aircraft to protect bombers from enemy fighters, and tankers to extend the fighters' reach. The Mountain Home wing would be made up of F-15 and F-16 fighter jets, B-52 bombers (later B-1Bs), and KC-135 aerial refueling tankers. All would train together to work as a unit.

The idea of a composite wing—also known as an "air intervention wing" and later renamed "air expeditionary wing"—was nothing new. The military assembled composite wings in China and Burma during World War II, during the Vietnam War, and in the Persian Gulf. The Air Force had known how to deploy a composite force since the 1950s. And the Navy had operated composite wings on carriers since the 1930s.[6] But once the hostilities were over, the aircraft returned to their own bases, which typically housed only one type of aircraft. The idea to assemble such a force during peacetime to train as a composite force came from Air Force Chief of Staff Gen. Merrell McPeak. It would provide rapid deployment to trouble spots around the world, he said. It would make a reality of the Air Force's idea of "global reach, global power" with a smaller force. And such a wing needed specialized training, which the state's proposed range would provide.

But before such a wing could be located at Mountain Home, the base would have to make it through the 1991 round of base closures. The Air Force told the commission that the existing Saylor Creek range would accommodate most of the wing's training, and the rest could be done at ranges in Utah and Nevada. The composite wing would not depend on the development of the state's range proposal, but the range would be an asset to the wing's long-term training, Air Force officials said. And the F-4s that had been slated for Mountain Home would go to the Idaho Air National Guard at Gowen Field in Boise.

On May 3, Andrus announced the beginning of the "public process" to establish the range, which would be a major part of changes planned by the Air Force at the Mountain Home base. The state would work with the Air Force on an environmental impact statement covering a new bombing range in Owyhee County. So far the plans had been developed in secret, but people would have plenty of opportunity to comment during the upcoming environmental impact analysis, Andrus said. Most people felt that by then it would be too late to affect the outcome. At the same time, the Air Force announced its intentions to complete an environmental impact statement in anticipation that the Base Closure Commission would approve the proposal to establish a composite wing at Mountain Home. The Air Force would consider the range and the wing, as well as proposed changes at the Idaho Air National Guard in Boise, as "a single joint venture" in a single impact statement.[7]

Though Air Force officials had assured the commission that the existing range would suffice, the "Description of the Proposed Action and Alternatives" dated May 24, 1991 said the existing Saylor Creek range lacked the capacity to provide the realism required for advanced training. The planning document said that without the state range, training would be limited to the Saylor Creek range "until another, more beneficial training alternative is identified." A June 10 version of the document noted that limitations of the range would require the use of simulators and out-of-state facilities.[8]

On May 31, Mountain Home appeared on the Base Closure Commission's list of thirty-six bases to be studied for closure or realignment. Mountain Home was included because of inadequate base housing, excessive distance to an adequate bombing range, and no low-level supersonic airspace. Community and elected leaders were tireless advocates during the fourteen public hearings on the bases slated for closure or realignment. The commission received more than 143,000 letters and took more than one hundred phone calls per day. Andrus and the Mountain Home Chamber of Commerce were no exception.

In advance of the commission's June 6 hearing on the Mountain Home base, Andrus and Mountain Home leaders wrote of Idaho's uncontested airspace, good weather, and low energy costs. And they pointed out that the state was working with the Air Force to develop expanded training capability, implying that the project was all but completed. U.S. Sens. Steve Symms and Larry Craig asserted that the base's electronic combat training capabilities—which only a year previous had been determined inadequate for the aging F-4s—were virtually unmatched in the Air Force.[9] And Craig said the commission, in its decision on the Mountain Home base, should consider the state's intent to expand training facilities—a proposal Craig and Symms claimed had broad-based support.[10]

Their combined efforts succeeded. On the day after the hearing, the commission struck Mountain Home from the closure list and approved the composite wing. That same day the Air Force announced a set of scoping hearings for an environmental impact statement on the realignment and the state's proposed range.

Skeptics were convinced that Mountain Home's appearance on the closure list was contrived. Herb Meyr noted that with access to a variety of ranges, plenty of open airspace, and no large cities nearby to encroach on that airspace, the Air Force was not likely to close the base. Still, officials later said that the state's promise of the range knocked the base off the closure list. Perhaps. But the senators were wrong about broad-based support for the range. Opposition included ranchers, environmentalists, hunters, hikers, river runners, bird watchers, private pilots, archaeologists, state and federal wildlife biologists, and American Indians.

These were not just a few environmental wackos, as supporters of the range tried to claim.

Once again realignment at Mountain Home was being recommended on the basis of a nonexistent range expansion. Once again an environmental impact statement on a proposed realignment at Mountain Home would be complicated by a proposed range expansion. And once again the coming planes would be used to try to leverage a proposed range that Air Force officials already had said was not needed. It was a ghostly echo of the Saylor Creek proposal, and the same logic, already publicly discredited, was heard again. The arguments were worded as if lifted directly from earlier documents.

A March 1991 Air Force memo noted of Andrus' proposed range, "Due to its remote location and stark environment, we are anticipating little opposition from local ranchers, Native Americans, or from the environmentalists."[11] But that attitude quickly changed. In May, Air Force Director of Environmental Programs Earnest O. Robbins said the Air Force had "a fundamental concern with linking the expanded range proposal with the (composite wing) in the EIS."[12] Robbins and others like him feared that if the two were combined, rising opposition to the proposal would delay the "beddown," or establishment of the composite wing.

Air Force officials moved quickly to try to distance the wing from the state range proposal. In a June 4 memo, a deputy chief of staff at Air Force headquarters expressed concern over "the apparent intent to link the governor's range proposal with beddown of the composite wing. Existing ranges are adequate to support training needs of the composite wing and other forces that will use them. Therefore, we should treat each initiative as an independent action for the purposes of an environmental impact analysis. Otherwise, we risk delaying the composite wing beddown due to unanticipated environmental complication with the range proposal."[13]

Air Force officials wanted the wing established on schedule. The Air Force's general counsel had said that while both should be covered, tying a decision on the range to the composite wing was not required. So the two were split, and Air Force officials began stating that expanding the range and establishing the new wing were separate actions not dependent on each other. As a result, the Air Force proposed another two-part impact statement—splitting the benefits of moving a new wing to Idaho from the environmental costs of providing expanded training for that wing. The first part would study the establishment of the composite wing and changes in airspace and at the Idaho Air National Guard, and it would evaluate the state's proposed range. If the range proposal were found environmentally and operationally suitable, a second impact statement would look at the environmental effects of such an expansion.

Toward the end of June 1991, with public scoping hearings on the first impact statement approaching, Gary Vest made another visit to Idaho, this time to threaten the state. Vest told The *Idaho Statesman* that without additional bombing and electronic combat ranges for the Mountain Home Air Force Base, the wing would not come to Idaho. Without the wing, the base might reappear on the Defense Department's closure list in two years.[14]

Then the BLM dropped a bomb of its own. On August 20, 1991, state BLM Director Delmar Vail expressed his concerns that Andrus' proposal included wilderness study areas. "*These areas can only be released for other uses through public law passed by Congress,*" Vail wrote to Gary Vest (emphasis in the original).[15] Because congressional action was pending on these areas, Vail asked the Air Force to consider two alternative areas southeast of the present Saylor Creek range. The move took Andrus by surprise. Worse, Andrus heard about the proposal not from the BLM but from the Air Force. He scolded the BLM for its "surprise." In a letter to BLM Director Cy Jamison, Andrus said, "I do not believe the decision by local BLM officials is in keeping with the spirit of cooperation and consultation that we thought we could expect."[16] During his visit to Idaho earlier that summer, Jamison had led Andrus to believe the BLM had no alternative proposals and that Jamison supported the state's proposal.

"I hope you will take a continuing personal involvement in seeing that this project is done properly, wisely and expeditiously," Andrus wrote.

Jamison replied, "The bombing range proposal is back on track. If I have anything to say about it…it will stay that way."[17]

The areas proposed by the BLM were more heavily used for livestock grazing, were closer to populated areas, and included unwilling sellers, Andrus said. The sites involved ten to thirteen grazing permit holders and several private landowners. Andrus was adamant about not forcing ranchers, which included J.R. Simplot, off the land. The state would not support the proposed BLM sites and would not propose the exchange of state land in those areas.

Not long after the BLM made its proposal, the range issue was moved from the Boise district to the state office. Vail said it was normal for sensitive issues to be handled by the state office. The flap over the BLM proposal had nothing to do with the move, he said.

In September 1991, writing to Jamison for the Idaho Conservation League, Mike Medberry commended Vail's proposal for describing the resources at risk—some of the most productive, valuable, and spectacular land in Idaho. Instead of trading it to the state, Medberry suggested the BLM consider the Big Springs area as "the heart of a National Conservation Area rather than the bull's-eye on a fighter pilot's screen." Medberry sent a copy to Andrus.[18]

Andrus' Air Force liaison Dave Jett responded on November 1, "We have always intended to meet the requirements of (the National Environmental Policy

Act)," including the need to consider alternative sites, "if the Air Force decides to pursue further study of the range proposal. Thanks to the BLM, the Air Force now has alternative sites to consider."[19]

But Andrus already had dismissed the BLM proposal, and it was never seriously considered.

The Air Force should have given more consideration to the BLM alternatives, the BLM's point man on the range proposal, Butch Peugh, said later. But at the time, the Air Force was adamant about not expanding the Saylor Creek Bombing Range. The BLM's purpose, however, was to make sure the impact statement was done right, and Peugh had wanted to make sure it said everything it was supposed to—and said it correctly. Peugh, who went to work for the agency during his summers while still in high school, rose through the organization and eventually became the area manager for the Boise Resource Area—the location of the state's proposed range.

The contractor writing the impact statement for the Air Force was Science Applications International Company of Boise. The company's bread and butter was the military, and it did whatever the Air Force wanted, Peugh said. Therefore, some effects were downplayed or glossed over. But the three most important issues were noise, wilderness, and bighorn sheep. Those were the areas the impact statement should have focused on, Peugh said. He was not against the range or the military, but he wanted the effects identified and acknowledged, not sugar-coated.

Through the summer of 1991, and in the lull after the demise of the Saylor Creek expansion proposal, Herb Meyr and Bob Stevens tried to keep the issue alive. The two had worked to get people outside of Idaho interested. Nobody in the rest of the country had heard of the governor's range proposal or knew about the canyonlands, so Meyr got the idea to put together a slide show. He wanted to show people in Idaho, Washington, D.C., and around the country that this was not the place for a bombing range. He received money from the Foundation for North American Wild Sheep and other environmental groups and approximately 240 slides from the BLM, Idaho Fish and Game, and individuals. Stevens took photographer David Stoecklein up over the canyons in his airplane to add some aerial shots. Mountain Visions in Boise helped put it all together with a narrator and music. Meyr transferred the slide show onto a videotape and sent copies to key officials. He also sent copies to national newspapers, including the *Wall Street Journal, Christian Science Monitor,* and *Washington Post,* along with Cable News Network and the Center for Defense Information. Most of them eventually published stories that helped bring some national attention to the issue.

Also during the summer, weekly meetings at the Wilderness Society office in Boise started up again. The group of opponents had re-formed as the Owyhee Canyonlands Coalition. A lawyer working for the Idaho Conservation League, Will Whelan, contacted a law-school friend who worked in the Boise office of the venerable Denver-based law firm of Holland and Hart. David Knotts recommended to Whelan that all the interests opposing the range form a legal entity; Knotts also said he would try to get his firm to represent the group pro bono. From that suggestion grew the Greater Owyhee Legal Defense (GOLD), which included conservation, hunter, and recreation groups. GOLD was officially launched on November 17, 1991. It was formed for the purpose of legal challenges but included many of the same people and groups as the Owyhee Canyonlands Coalition. Most of the work representing the group fell to Murray Feldman, a recent law school graduate who had come to work for Holland and Hart's Boise office in January 1990. By the time GOLD formed, he already was hard at work on the issue.

In his review of the Air Force's draft impact statement on the composite wing, released in October 1991, Feldman noted that the document did not consider a reasonable range of alternatives to the state's range. Air Force officials said the decision at hand was limited to the operational and environmental suitability of the state's proposed Big Springs Training Range. Feldman said that the suitability could not be determined without evaluating feasible alternatives.

In January 1992, the Air Force released the final environmental impact statement, and the decision to establish the composite wing at Mountain Home followed in March. No one was surprised that the decision included pursuing the state's range proposal.

The Air Force had plans for the state's 160,000-acre proposed range. Officials wanted a practice bombing range that could be approached from all sides. The existing Saylor Creek Bombing Range was located in the northeast corner of the military airspace over Owyhee County. It could be approached only from the south. Plans for the range called for at least two bomb impact areas that would make up about 25,000 acres—the rest would provide a buffer. The Air Force planned to drop about twenty-four thousand practice bombs—up to two thousand pounds each—on the impact areas within the proposed range annually. Nearly all of those dummy bombs would use spotting charges, which released a small puff of smoke to score bombing runs. The proposal also would expand the use of flares and chaff throughout the entire military operating area—they had been allowed over only the Saylor Creek range. Because the spotting charges and flares would increase fire risk in the dry, trackless desert, the proposal included a 150-foot wide firebreak around targets.

The proposal included more than thirty sites—from .5 to 5 acres each—for mobile radar emitters that would simulate enemy air defenses. The control center for the electronic combat range already sat on a rise just off of Idaho Highway 51 at Grasmere. The emitters would be located mostly along existing roads, including Mud Flat Road, the Bruneau-Three Creek Road, and dirt roads off Highway 51. The proposal also included supersonic flight higher than ten thousand feet above the ground over most of southern Owyhee County, creating an average of about ten sonic booms a day or approximately three hundred a month. To reduce potential effects on wildlife where the range would border the canyons, Fish and Game suggested a three-mile buffer between the range and the canyon rims.

"We're going to have the biggest low-level training area anywhere in the United States," Bob Stevens predicted. "Governor Andrus will wish long after he's left office that he never went to bed with the Air Force."[20]

Chapter 6 / Split Range

L ONG BEFORE THE FINAL IMPACT statement on the composite wing was released, a split developed between Cecil Andrus and the Idaho Department of Fish and Game which never quite healed. The department's concerns over "potentially serious impacts to fish and wildlife resources" were summarized in a December 6, 1991, letter from Fish and Game Commission Chair Norm Guth. The letter suggested that better places for the range could be found and accepted Andrus' offer to allow Fish and Game to work with the governor's staff to find those places that would satisfy the Air Force and reduce the effects on wildlife. But when word of Fish and Game's reservations about the range showed up in the news, Andrus blew up. He accused Fish and Game employees of leaking to the media that department officials had expressed doubts about the state's proposal and its effects on wildlife and suggested that there may be better places for a range.

"I continue to be frustrated by public statements made by Fish and Game Department employees," Andrus wrote to the Fish and Game commissioners. If those employees think there is a better location for the range, "for God's sake, let them tell us where." He suggested they should name that location and give assurances that landowners would sell—if they knew such a place existed.[1] But it was not the job of Fish and Game employees to propose alternatives, only to make suggestions and evaluate them, Fish and Game Director Jerry Conley countered.

Conley never openly opposed the range, but at the risk of his career, he did little to make the range proposal work. He deftly handled the volatile political situation and allowed his staff of biologists to do their jobs. The result was that verifiable scientific data surfaced and cast doubts on the proposal. The closer Fish and Game looked at the high desert range, the worse the proposal looked.

Public criticism eventually forced Andrus to modify the proposal. Though the change looked like an improvement, it also focused public attention on a little-known area. More and more people became aware of the resources the area offered, resources they increasingly feared would be jeopardized by the proposal.

In January 1992, Conley and Andrus began discussing with Air Force officials the idea of splitting the range to put one half on the north side of the Owyhee River's East Fork and another to the south. And with reluctant agreement, officials from Fish and Game, the Bureau of Land Management, the Air Force, and the governor's office started meeting quietly to see whether there was some way of changing the boundaries of the range. Again the public was cut out of the discussions. State officials said the issue was too sensitive to talk about publicly. Fish and Game spent the first few months of 1992 poring over maps,

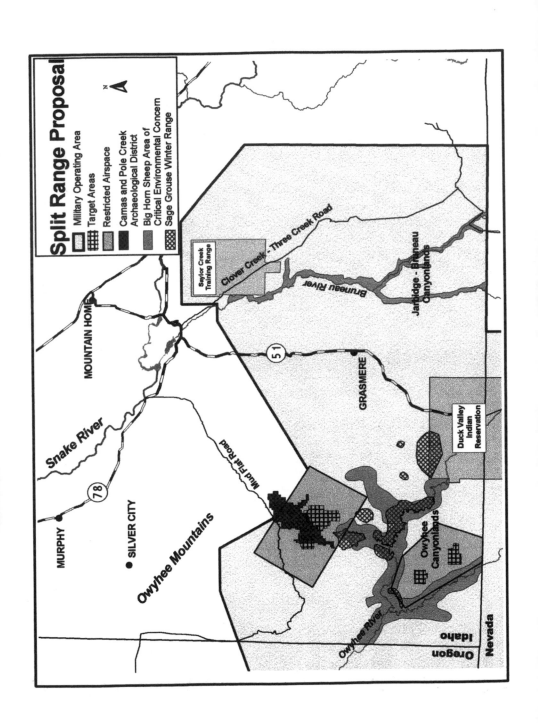

Split Range Proposal

Legend:
- Military Operating Area
- Target Areas
- Restricted Airspace
- Camas and Pole Creek Archaeological District
- Big Horn Sheep Area of Critical Environmental Concern
- Sage Grouse Winter Range

MURPHY
78

MOUNTAIN HOME

Snake River

SILVER CITY

Owyhee Mountains

Mud Flat Road

Saylor Creek Training Range

Clover Creek - Three Creek Road

Bruneau River

51

GRASMERE

Jarbidge - Bruneau Canyonlands

Duck Valley Indian Reservation

Owyhee Canyonlands

Owyhee River

Oregon
Nevada
Idaho

looking for a place to put a bombing range that would be acceptable to most people, Conley said.

Fish and Game biologists had criticized the range proposal because it failed to identify and analyze clearly foreseeable problems. The canyon of the Owyhee River's East Fork forms a natural barrier to wildlife concentrations north of the river. If habitat were lost to range development, fire, and human intrusion, animals might be unable to migrate south to other suitable habitat.[2]

BLM officials had their own concerns about the Big Springs range. The area was one of the few remaining in Idaho where native vegetation still dominated. It included all or parts of four wilderness study areas—more than one third of the proposed range was under study as wilderness by the BLM. And the agency was bound by law to manage such land as though it were wilderness. The proposed range contained:

- 58,700 acres of wilderness study areas.
- 11,000 acres of critical mule deer winter range.
- 21,000 acres of prime antelope winter range.
- More than 22,000 acres of sage grouse wintering areas.
- Forty-four miles of proposed Wild and Scenic Rivers.
- Fifteen river miles of bald eagle winter habitat.

Mule deer wintered in the southern third of the proposed range, and sage grouse used seventy-five percent of it as nesting habitat; bighorn sheep used the canyons and antelope the entire range. The quiet, solitude, rafting, canoeing, and hunting that drew people to these areas also would be affected. All of this would be subjected to low-flying, high-speed jets, sonic booms, and chaff and burning flares dropping from the sky.

BLM officials said it was inappropriate to conclude that military overflights would not be a factor affecting a wilderness designation. While overflights do not rule out such a designation, they do diminish the area's wilderness value and may affect congressional consideration of such designations. As Rick Bass states in his book *The Lost Grizzlies*, what constitutes wilderness legally—morally—is that the area have no motorized vehicles. Is it truly wild if machines can hover and flit and roar only a few feet off the ground?[3]

On April 2, 1992, Governor Andrus announced a modified version of the range proposal. Instead of 160,000 acres north of the Owyhee River's East Fork, the range would be split. The new proposal would consist of approximately 96,000 acres north of the canyon and 70,000 acres south of it. Within those two areas, the state hoped to consolidate 25,000 acres—a portion of that on each side of the river—for bomb impact areas through trades with the BLM. The Air Force would use the three ranges—the two new ones and the existing Saylor Creek range, along with an electronic battlefield in-between.

The revised proposal would resolve issues raised by wildlife experts, Andrus said. It would provide expanded training facilities without compromising the state's environmental values. Though Andrus had assured the Air Force of the full support of the Fish and Game Commission, that support was grudging and conditioned. "If we had our druthers, there'd be no range," said Fish and Game Commission Chair Richard Meiers of Eagle.[4] That sentiment was echoed by Commissioner Keith Carlson, who recognized that an expansion of the present training capacity was likely and the split version would be less harmful to wildlife. The split-range proposal included the same conditions as the Big Springs proposal—no live ammunition and no supersonic flight lower than ten thousand feet above the ground. Fish and Game again asked for a three-mile buffer zone from key wildlife areas.

Air Force officials approved the changes and said they would be included in the upcoming draft environmental impact statement. On April 24, the first round of public hearings on the state's proposed range were announced.

The proposal was met with skepticism. Hunters still were concerned about access and the health of the bighorn sheep herds in the canyons. Environmentalists were concerned about the desert ecosystem and the potential loss of outdoor recreation opportunities. The Owyhee Cattlemen's Association criticized this proposal for being developed in secret like the previous one, and for ignoring the willing seller approach to selecting a site. Though the south half of the range would not include any private land, it would wipe out significant portions of at least one grazing allotment, and it would limit access to the 45 Ranch on the South Fork of the Owyhee River. Only the effects on wildlife and the needs of the military were considered in the selection, which gave no consideration to the effects on the people who lived and worked in the area, the cattlemen said.[5] The decision to bring the composite wing to Idaho was made after the 1990 Defense Department's moratorium on land acquisitions and without assurance of a state range, the cattlemen noted. With a $4.4 trillion national debt, it did not seem prudent to continue an expensive review process for a range that might not be necessary.

The following month, on May 8, 1992, the Greater Owyhee Legal Defense filed a lawsuit in U.S. District Court in Boise challenging the Air Force decision that established the composite wing and the decision that the state's bombing range proposal was operationally and environmentally suitable. The Shoshone-Paiutes, urged by Bob Stevens, filed their own suit. And rancher Dick Owen and his wife Marie, who owned the Big Springs Ranch, filed a third—all on the same day.

The lawsuits said the Air Force's impact statement—upon which the decision was based—was inadequate and violated federal environmental law. The document did not consider an adequate range of alternatives, adequately evaluate the environmental effects of the proposal, or adequately discuss mitigation for the adverse effects. And the wing and the range should have been considered in the same impact statement. GOLD had submitted detailed comments on December 10, 1991, pointing out numerous deficiencies in the draft impact statement. The final impact statement was printed December 17, "obviously too soon for the Air Force to have seriously considered the comments of the public." The Air Force did nothing to respond to or correct the deficiencies before issuing the final decision, forcing GOLD, the Owens, and the Shoshone-Paiutes to seek judicial review, they said.[6]

Supporters of the range railed against the lawsuits, calling them antimilitary and unpatriotic. The issue, however, was not one of patriotism, but of requiring the Air Force to abide by the laws of the land, GOLD attorney Murray Feldman said. The suits left a cloud of uncertainty hanging over the state's range proposal.

Despite the lawsuits, however, Air Force officials decided to go ahead with preliminary public hearings in mid-June 1992 on the impact statement that would cover the state's range proposal. But they flew headlong into a public relations disaster with the format they chose for those hearings. Normally at public hearings people are seated in a meeting room and those who wish to testify in front of the entire audience. But for the first round of hearings, Air Force officials, apparently tired of the frontal assault of past hearings, tried a different tack. They set up booths with photographs, maps, and diagrams pinned neatly to large bulletin boards. It looked like a public relations campaign or a trade show. Many people objected. Some said the Air Force was manipulating the process to avoid public scrutiny of its proposal.

In the first booth, a video with Governor Andrus and several Air Force and National Guard officers promoted the range proposal. At other booths, well-scrubbed airmen in dark blue, creased trousers, and shiny black shoes eagerly answered questions. But they knew little beyond the obvious, and they bucked tough questions up the chain of command. Specific information on the proposal was not available, making meaningful public involvement difficult. When one man asked a master sergeant about the proposed emitter sites, he was told their purpose was to simulate surface-to-air missiles, anti-aircraft artillery, and surveillance radar. But when he asked about the number of such sites and their location, the sergeant responded that he was not authorized to provide that information. The booths may have provided more information about the proposal, but the information was selective and the format intimidating.

In another booth people could offer their testimony to an Air Force officer and a court reporter. It was less intimidating, officials said. But one man likened it to a confessional, and anyone who wanted to comment had to face the stern-faced Air Force alone.[7]

Though the format did not violate anyone's right to speak their mind, it certainly did stifle the exchange of information and ideas that normally occurs in a public hearing. The format was blasted in southern Idaho newspaper editorials. One admonished Air Force generals to muster their courage "and look the taxpayers in the eye."[8] The Twin Falls *Times-News* also responded with "Operation

Sound Barrier." The hearing format prevented formation of any community consensus, the newspaper said. But if the Air Force would not provide a public forum, then the paper would. The letters poured in. In less than two weeks, more than fifty people wrote in, all but half a dozen criticizing the range proposal. The letters showed that people were unhappy with the meeting format as well as the range, and they were still looking for the Air Force or the governor to show a need for it.

At most of the hearings, opponents of the range proposal set up their own booths in the hallways outside the meeting rooms. These booths presented other points of view, including maps, pictures, and Herb Meyr's slide show. And people were on hand to explain the process, answer questions, and help others comment on the proposal. They provided the public with the rest of the story, said Brian Goller of the Owyhee Canyonlands Coalition. Air Force officials and others attempted to limit the amount of information available on the proposal, giving an incomplete picture of the full range of impacts. "We won't let them get away with that," Goller said.[9]

Meanwhile the support Andrus counted on was eroding. Fish and Game Commissioner Richard Meiers' response to the hearings was that he thought the range was a done deal. He had no concrete evidence upon which to base the feeling, but he thought the decision had already been made because Governor Andrus and Air Force brass were not seriously considering other sites. Air Force officials countered that they were considering all the alternatives, including expansion near the existing Saylor Creek Bombing Range and no new range at all.

Meiers did not believe it.[10] Neither did many other southern Idaho residents.

Though the range was supposed to be owned and operated by the state, the Air Force had agreed to pay for about sixty-two hundred acres of private land the state would be obligated to buy. These private lands would be part of the mitigation for any effects the range might have—a bonus for recreation, state and Air Force officials said. The state Parks and Recreation and Fish and Game departments would manage the areas. Some critics saw it as a shallow effort to buy the cooperation of the two agencies; Owyhee County already had tremendous recreational opportunities and plenty of wildlife habitat.

The U.S. Army Corps of Engineers, which conducts real estate appraisals for the military, in October 1991 had analyzed the proposed real estate transactions and recommended against construction of the range. Bighorn sheep were one of the reasons. Andrus was told of the decision October 10. On October 27, he instructed Parks and Recreation to transfer money from its 1994 budget for the City of Rocks National Reserve in Cassia County to development in Owyhee County. In November, Andrus told Air Force Secretary Donald Rice that the Corps did not like the proposal much. Rice assured the governor that the Air

Force would go ahead with the appraisals. Earlier that year, the Air Force had questioned whether it could legally buy the sixty-two hundred acres, estimated at that time to be worth $3 million to $4 million. The Defense Department's 1990 moratorium covered any purchase, withdrawal, lease, permit, or use agreement involving more than one thousand acres or a purchase price of more than $1 million.

The proposal, however, was attractive to Air Force leaders. It would increase training capability in Idaho at a fraction of the cost normally required to acquire a bombing range. By acquiring private lands, the Air Force could control access to the bomb impact areas on public lands, limiting potential future incompatible land uses and meeting federal requirements to control lands under restricted airspace.[11]

One of those private landowners, Robert Nahas, hoped to sell his twenty-five-hundred-acre ranch for $1.75 million. A 1991 appraisal had set the value at about $500 per acre, or $1.25 million, and $150 per animal-unit-month for the grazing permits. The Owyhee County assessor set the value of the Nahas ranch at $1 million. In June 1991, Nahas had proposed trading with the state for fifty acres of prime recreational property on Payette Lake near the resort town of McCall worth about $3 million. Nahas would turn over his Deep Creek Ranch and $1.25 million in cash for the lakefront resort property. The state in turn could sell the ranch to the Air Force, Nahas proposed. The Department of Lands turned him down, however, saying the property on Payette Lake was worth $3.5 million and taking more than half the purchase price in cash would constitute a sale and as such required the state to follow public sale procedures.

The other private landowners, Joe Black and Sons, had agreed to sell approximately thirty-seven hundred acres and the leases to approximately ninety-five thousand acres of BLM grazing land worth a similar amount. The Air Force was expecting to spend about $6.7 million.

Meanwhile, the landowners were getting antsy about earnest money. In June of 1991, Black and Nahas had written Andrus, saying it was unreasonable for them to wait indefinitely without something in writing. Andrus bucked the request up to Gary Vest for "a satisfactory solution to the earnest money issue."[12] For the state to request the exchange without any written agreement with the private landholders would be risky, Andrus said. Vest replied that the Air Force would be pleased to discuss the earnest money.[13]

That was not good enough for Andrus. The earnest money issue had to be settled quickly to maintain the state's credibility with the ranchers, who were key to the range proposal, he told Vest. He also wrote of his concern to Rice, and the correspondence continued through the fall of 1991. But the Air Force continued to evade direct answers, saying no money could change hands until the environmental impact process was completed. Until then the Air Force would not be able to determine what property actually would be needed, Vest wrote. Putting money on a piece of land would seem to prejudice the outcome of the environmental impact process.

Nahas had written Rice, in November 1991, wondering where his cows would graze the next season. He got basically the same reply as Andrus—wait. Rice replied cordially that Nahas need not worry about where his cows would graze. "I am certain you can assure your cows they will have familiar pastures through 1993," Rice wrote, saying also that the purchase of private land would be completed along with the land exchange as soon as the impact statement on the range was completed. And that process would be completed by late 1993.[14] Nahas eventually got tired of waiting and withdrew his offer to sell his ranch. But Andrus and other officials, who earlier had insisted they needed the ranch for the range, said the split-range proposal would not be affected by Nahas' withdrawal.

In early November 1992, state officials announced their intention to swap nineteen thousand acres of state land for twenty-one thousand acres of federal property to create the range. The announcement was the first public indication that the state was actively trying to arrange a deal to create the range. Trading real estate with the BLM was key to the state's range proposal. The state Land Board already had approved a process for the land exchange back in December 1991. But the actual exchange still was a long way off.

The environmental impact statement on the range, when it was released, would include information on the proposed land trade, skeptics were told. But the critics feared that by the time the process got that far, any meaningful public participation would be too late. Locking in the land swap before the impact statement was completed would prejudice the outcome. It meant that the Air Force would not take a serious look at other sites—such as those proposed by the BLM—or at ways to do more training at ranges in nearby states. And once the swap was completed, the Air Force or the Idaho Air National Guard would be able to start construction on the range. But Dave Jett said that the exchange would be done exactly as any other land exchange between the state and BLM once the Air Force had made a final decision on the range. It would be up to the BLM to conduct any additional public hearings that might be necessary.

The land swap would give the BLM state-owned parcels scattered throughout southwestern Idaho, helping the agency consolidate some of its holdings. In exchange, Idaho would get enough federal land to create the two proposed bomb impact sites for the split-range proposal. But the existence of Wilderness Study Areas continued to shadow the project—congressional approval would be needed to include those areas in any swap. A portion of the land would require protection and mitigation plans for archaeological areas, Craig Gehrke of The Wilderness Society told the Land Board. It would be the state's responsibility to carry out such plans, and that could be costly, he said. Other considerations would include endangered species and wetlands. He suggested the board complete a thorough cost analysis before it approved any land trade.

On December 4, 1992, before it was clear just who would pay for and own the property, state officials divided up the private parcels in Owyhee County—a small part to the Department of Lands and the rest split between Parks and Recreation and Fish and Game. Everyone still thought the Air Force would buy the land and turn it over to the state. Not until the following year would the public learn that the state would not own the range after all. After a December 1993 meeting, it appeared the Air Force would own at least part of the range. And it appeared that Idahoans would have little to say about how it was operated. The effects on wildlife and public access would be addressed in a range management plan to be developed later. A draft outline of the plan would be available at the public hearings. But there would be no opportunity for the public to be involved, or to comment.

Meanwhile, the 1992 election brought a change in power. Voters had swept out the George Bush administration and installed President Bill Clinton and Vice President Al Gore, who were seen as friendly to environmental issues. Andrus' Democratic Party now controlled the White House, and Andrus was an even more powerful and influential politician. In Idaho, however, Republicans gained more strength than they previously had. Boise Mayor Dirk Kempthorne had replaced Steve Symms in the U.S. Senate, and state Sen. Mike Crapo had taken Democrat Richard Stallings' position in the House. Two years earlier, Rep. Larry Craig had moved into Sen. Jim McClure's position. Idaho's mostly Republican congressional delegation backed Andrus' range proposal. And that spelled doom to most range opponents. Many were convinced the range was a done deal.

But after Clinton took over the White House in January 1993, a few Greater Owyhee Legal Defense members presented a proposal they had developed over the winter. Murray Feldman and former Natural Resources Defense Council lawyer Eric Christensen urged President Clinton to issue an executive order that would include a moratorium on all military expansion until the Department of Defense could assess national training needs. At the time, the military had proposed withdrawing or otherwise acquiring 4.6 million acres of public lands in California, Colorado, Idaho, Montana, Nevada, Utah, and Washington.

The executive order would close the loophole that allowed the National Guard to withdraw large amounts of public lands for military training without congressional review. The proposed moratorium would force military planners to assess their current training needs before seeking additional land and protect public land resources until an assessment clearly showed the need for withdrawals. The assessment should be done by a blue-ribbon panel that would include nonmilitary experts, and it should provide opportunities for public participation. It would require all-but-nonexistent military interbranch coordination and planning to make better use of existing facilities.

Congressman Bruce Vento of Minnesota introduced and held hearings on a bill that was similar and would have changed the way public lands and associated airspace could be withdrawn for military use. It was not enacted.

Other members of Congress, environmental groups, federal land management agencies, and the General Accounting Office had raised the issue of assessing military training needs before spending tax money for new facilities that might not be needed. The GAO already had called for better planning and interagency cooperation in obtaining military withdrawals of land and airspace.

Clinton did not act on the proposal.

Chapter 7 / Range Criticized

THE DRONE OF A single-engine plane cut through the haze early one morning in July 1993 before the plane landed at a small, rural airport used mostly by local crop dusters. Without cutting the engine, Bob Stevens opened the door and handed out a brown paper package that rattled in the prop wash.

"Remember, you don't know who sent you this," he said and winked. "Good luck."

Stevens was only a courier. He grinned, slammed the door, revved the engine, and turned back toward the runway. Moments later the light plane was airborne again. Inside the package was an unofficial preview copy of the environmental impact statement on the state's proposed bombing range. The Air Force, the BLM, and the state of Idaho had refused to release it. But the illicit copy was made public, undermining an ill-fated effort to keep key information about the proposed range out of the hands of the Idaho Department of Fish and Game.

The preliminary document laid bare many of the shortcomings of the proposal, giving critics a clearer target, and was an embarrassment to Air Force officials who earlier had said the draft environmental impact statement would be released for public comment by May or June. The document's official release was delayed by ongoing internal disagreement and issues that needed to be resolved—issues officials remained tight-lipped about. And opposition to the range proposal was growing in Washington, D.C., giving hope to the opponents in Idaho.

Airing the illicit preliminary document was also somewhat of an embarrassment to the governor's office. Earlier that year, the Air Force had issued copies of the preliminary draft to selected agencies for comments and corrections—standard procedure. Fish and Game, as a cooperating agency, however, received only a few chapters, despite repeated requests for a review copy. When asked, Governor Andrus' liaison Dave Jett said that was because the entire document had not been printed yet. But when confronted with the complete illicit copy, he admitted that the first two chapters were not given to Fish and Game. Those chapters described what the Air Force planned to do, where targets would be located, and how the range would be operated. Jett said the chapters were withheld because those plans were constantly changing and not complete. And he was convinced Fish and Game officials would use the information to support their contention that the range would wipe out the bighorn sheep. The governor's office also wanted Fish and Game kept out the planning process because of previous leaks of sensitive information to the public. Fish and Game, however, said the leaks came from Jett's office.

Bob Stevens pilots his Cessna 185. *Photo by the author.*

Without the key chapters, Fish and Game could not determine what effects the range might have on the bighorn and other wildlife in the canyons between the two halves of the range. Biologists had predicted a potential for serious effects on the sheep, but department officials denied that, saying the range would wipe out the sheep. Fish and Game officials had helped develop the governor's split-range proposal, thinking it would be a way to reduce negative effects on wildlife in the canyons. But departmental support for the proposal had been contingent on restrictions that included a three-mile low-level overflight buffer between the target areas and bighorn lambing habitat and avoidance of wildlife areas during lambing, fawning, and nesting seasons.

Jett said those conditions were only recommendations and that no previous agreement had been made on airspace and the three-mile buffer. But in a letter dated August 25, 1993, Andrus reiterated the state's restrictions. And he had repeatedly told people in Idaho that the range would be located three miles away from the canyon rims, avoiding sensitive wildlife areas, and that it "would not dramatically or adversely affect the environment on the ground."[1]

The preliminary draft document did not include those limits, and the critical wildlife areas had not been identified as agreed, Fish and Game officials complained. One proposed restricted target area overlapped habitat identified as critical for bighorn sheep. Another included areas along Dickshooter Creek known to be favored by ewes and their lambs. No restrictions on altitude or flights below canyon rims were included, nor were public access issues. The document also proposed the use of live 20 mm ammunition in strafing runs on both halves of the range, despite assurances that live ordnance would not be used.

BLM officials in Boise complained that they were not given time for much more than a cursory look at the preliminary document. Officials had expected thirty days; the Air Force gave them only ten working days. The proposal in the preliminary draft included several areas under consideration by BLM for wilderness designation and about two hundred miles of streams and rivers under study for potential wild and scenic river status. Nearly half the proposed flights in the area would be less than one thousand feet above the ground. While overflights of potential wilderness areas were not expected to increase significantly, more of those flights would be at low altitude. BLM scientists wanted the effects of these low-level overflights on wilderness study areas and on wild and scenic rivers, as well as other discrepancies with the state's recommendations, covered more fully in the impact statement.[2]

BLM officials had expected more of a role than just being told what decision the Air Force had made, state BLM Director Delmar Vail said.[3] The agency already felt pinched between Air Force critics and its responsibilities as manager of the land the state wanted to trade to establish the range. But Vail took his task seriously, affirming that any land exchange would include reasonable alternatives to the state's proposal and full public participation. He would not be a pushover, and he was by no means a patsy for the Air Force.

When BLM officials pointed out flaws in the impact statement, they were accused of being uncooperative by the Mountain Home Chamber of Commerce Military Affairs Committee. BLM officials were trying to force restrictions on overflights of potential wilderness areas, committee president Mike Miller said. The Mountain Home newspaper accused BLM Boise District Manager Dave Brunner of throwing up as many roadblocks as he could think of while saying he supported the range. The committee urged people to write letters in support of the proposed range to the head of the BLM, the Secretary of the Interior, the governor's office, and Idaho's congressional delegation. Dave Jett, who served on the committee's board of directors, offered to supply envelopes and stamps. He said his participation in the letter-writing effort had no connection to his job as the governor's liaison.

Other supporters chipped in as well. During the summer of 1993, Andrus taped a television spot encouraging people to become involved in the public hearings and comment period that would follow the public release of the impact statement. The Idaho Alliance, Inc. of Eagle paid for airtime for the spot. The nonprofit group was formed to provide information about the Mountain Home Air Force Base and the proposed state range that was needed "to preserve the base as vital to our economy and our nation's defense needs," said Grant L. Petersen Sr., a Mountain Home car dealer.[4] The alliance spent about $200,000 promoting the range, saying it was concerned about half-truths and inaccuracies circulating about the proposal.

Meanwhile, in an effort to resolve the problems with the preliminary draft—already aired in the media—the Air Force in July flew Idaho and BLM officials to Langley Air Force Base in Virginia to discuss the lagging schedule. Officials would not comment on details of the meeting but said it covered the study and preservation of cultural resources, bighorn sheep, and the proposed land exchange. They also talked about alternatives to the proposal and the wilderness study areas within the range.

At that same meeting Air Force leaders dropped their request for a memorandum of understanding that would have given them ground control of about three million acres in the range's "Region of Influence." Instead the Air Force would assume control through a Range Management Plan to be completed after the range was approved. The plan would spell out just how the range would be operated and what precautions taken to minimize effects on wildlife, but it would not be subject to public comment or congressional approval. It was not clear at the time exactly who would control the range. Hints were strong that it would be the Air Force, despite the governor's assurances to Fish and Game and the public that the state would retain control. After the meeting, the Air Force said it expected to release the draft impact statement in early September, with a public comment period to run until December 10, 1993.

In August, BLM Director Jim Baca paid a fateful visit to Idaho. Baca, appointed in May 1993, was seen as friendly by environmentalists for his tough stand on mining and grazing as public lands director in his home state of New Mexico. And the land exchange would have to go through Baca as head of the BLM. Bob Stevens had made sure Baca was well briefed on the range issue before he started in his new job. Range opponents invited him to see the proposed range and the land his agency would be trading away to the state.

Andrus, who had met with Baca in Arizona earlier in 1993, knew Baca would be hearing about the range proposal from opponents, and he also invited Baca to have a look. Baca agreed. He arrived in Idaho in late August 1993 to oversee a land exchange in Coeur d'Alene and the following day flew over the proposed range with a group of BLM employees. Andrus had asked to go along to correct any incorrect impressions Baca might have or might get about the range. But he was told that only technical people would be going. Baca instead agreed to meet later with Andrus. But Baca scheduled a news conference of his own just before their meeting.

Andrus caught wind of the news conference and, suspecting a palace coup, slipped into the back of the room. He learned that members of the Foundation for North American Wild Sheep—vocal opponents of the range proposal—had flown with Baca, as had Dave Brunner, the Boise District BLM supervisor already labeled by Andrus as an opponent of the range. Baca said he was skeptical of claims that low-flying jets would not affect the area's bighorn sheep herd. And he was worried that practice bombs would start wildfires in that trackless desert, priming the area for a massive invasion of cheatgrass and other noxious weeds. The range issue should be decided by science, not politics. Other powder-keg issues, such as the spotted owl controversy in the Pacific Northwest old-growth forests, blew up because of political considerations, not scientific ones, Baca said.[5]

But his biggest mistake, as far as Andrus was concerned, was calling the proposal a bombing range. Andrus later said he disliked the tone of Baca's voice when he called it a bombing range. And Baca may have been confused. At one point during the press conference, he referred to live ordnance. As a former Air Force staff sergeant, Baca said, he had helped put out fires and supervised crews that cleaned up bomb fragments at ranges where practice and live bombs were dropped. He was familiar with the Air Force, and he knew what a bombing range looked like. Baca thought the area was special and would be better preserved as wilderness.

Andrus said Baca's characterization of the proposed range left the impression of exploding bombs. He maintained his proposal was more properly termed a "training range." Bombs dropped on the proposed range would not contain any explosives, Andrus claimed, only talcum powder, to mark the spot of impact.

They were both wrong.

No live ordnance would be used on the range, but the dummy bombs would include two types of explosive spotting charges: the Hot Spot, which uses red phosphorus to produce a brilliant flash of light and dense white smoke, burns at 2,732 degrees F, and produces a six- to eight-foot flame easily capable of setting fire to dry desert vegetation; and the Cold Spot, which uses a small gunpowder charge to crush and discharge an ampoule of titanium tetrachloride. The chemical creates a white smoke cloud when it reacts with moisture in the air, though it is not generally capable of starting a fire.

When Andrus and Baca finally met, Andrus erupted at him, Baca said later. He thought Andrus was out of control. But Baca stood his ground. Andrus, in a letter to Interior Secretary Bruce Babbitt, later lamented that he, a friend and supporter of President Clinton and Babbitt, found himself locked in conflict with the administration over the proposal—before the draft impact statement had been released. Babbitt did not have "enough political allies in the western United States to treat us this shabbily," Andrus said. Baca had made up his mind about the proposal without ever giving Andrus the courtesy of hearing his side of the issue and without spelling out his concern to Andrus before condemning the range proposal in public.[6]

Baca disagreed. He and Andrus had spoken before the news conference. Baca maintained that he had only repeated the comments of BLM employees, who said the agency should take a second look. "And Andrus went ballistic," he said.[7]

Range opponents jumped to Baca's defense. That November, the Foundation for North American Wild Sheep named Baca its Outstanding Federal Statesman for 1994, in recognition of his stance on the bombing range. To many, the references to twenty-five-pound dummy bombs brought to mind men in goggles and leather jackets, white scarves fluttering, dropping little bombs out of old biplanes. The actual proposal, however, would have been a sophisticated electronic combat range. Herb Meyr termed it a high-threat, supersonic battle-field. Bob Stevens said it would have sterilized the ground from the air.

In September 1993, range foes were joined by an unexpected ally when Sen. Harry Reid, a Nevada Democrat, asked the Senate to halt the "unneeded Idaho Training Range." Reid said he was concerned about the effects of noise from military jets on people living at the Duck Valley Indian Reservation, which straddles the Idaho-Nevada border. No other group of people would be as af-fected by the proposed range. "So the questions remain: Why are we going to spend money on an unneeded military installation? Why are we going to take more away from the Shoshone-Paiute Tribes? Why are we going to desecrate lands sacred to a people who have already been taken advantage of?"[8]

Sen. Dirk Kempthorne of Idaho, a Republican, intervened on behalf of the Air Force, and the two senators agreed to a compromise, requiring the Air Force to study the feasibility of creating a fifteen-thousand-foot ceiling and a fifteen-mile buffer around the reservation.

In late September, Congress delayed a Pentagon request for $6.7 million to buy the private land for the proposed range until uncertainty about the need was settled. The Senate blocked any money for range expansion or land acquisition until Defense Secretary Les Aspin provided written certification "that the project is required for training and readiness." Officials insisted the money would be used to buy the ranchers' property only if the Air Force selected the state's pro-posed range. But until an impact statement on the range proposal was released, getting the money would be premature, the Senate noted.

Reid was not the only member of Congress who would oppose the state's range proposal. One afternoon in October 1993, during one of his many trips to Washington, D.C., Rick Johnson of the Sierra Club's Seattle office became caught in a sudden downpour. He and fellow traveler Brian Goller of the Idaho Conservation League sought shelter in a portico at the back of the Capitol. Also seeking shelter there was a well-dressed gentleman who looked familiar to Johnson. They exchanged greetings the way folks do when chance throws them together. The gentleman turned out to be Congressman Peter DiFazio of Or-egon, a man Johnson had dealt with many times over the spotted owl issue. Johnson introduced Goller, who then related the story about the bombing range.

That meeting ultimately led DiFazio to write a letter to Interior Secretary Bruce Babbitt. The letter, signed also by the rest of the Oregon delegation, ques-tioned the proposal for a state-owned range and urged Babbitt to reject it. In-creased aircraft activity associated with the range would extend into eastern Or-egon. The five Oregon representatives were concerned that the need for the range would not be discussed in Congress, but the range would affect the solitude and quiet in the Owyhee Wild and Scenic River, Steens Mountain, and other remote areas of southeastern Oregon. The letter was the first solid congressional opposition to the state range proposal.

While he was in Washington, Johnson had arranged to meet with Air Force Secretary Sheila Widnall to discuss the range. They were to meet at the Pentagon with Washington officials and lobbyists and link by telephone conference call to Craig Gehrke, Bob DiGrazia, and Herb Meyr, all in Idaho. Johnson was brought by a convoluted route, moving ever inward through the levels of the Pentagon, to finally arrive in the meeting room, he said. Around a large table sat many political appointees and D.C. conservationists surrounded by men and women in uniform, silent except to operate the communication equipment. But the equipment did not work. So the whole group moved to another room with similar equipment. It did not work either. Widnall appeared irritated and a little embarrassed, Johnson said. Finally a speakerphone was brought in. It did not work either. The meeting went on without the Idaho activists, and without their knowledge of the details of the issue. But it was then that Johnson started thinking they could win.

Meanwhile, Goller was in Washington to visit with Idaho state Congressman Larry LaRocco, and he went around to the D.C. offices of other environmental groups to talk about the bombing range issue. He was surprised at how many people had heard about the proposal. The issue was beginning to get some national attention. The Sierra Club had paid for his ticket and put him up in an apartment on Capitol Hill. People began to realize that the issue was important enough to send someone from Idaho all the way to D.C., and that gave the issue credibility with many people in a town full of skeptics.

Finally, in November 1993, the Air Force released its environmental analysis of the state range proposal—six months late and disappointing to many. For the $3.1 million the Air Force spent, Fish and Game commissioners Keith Carlson and Richard Hansen thought it was poorly written and lacking in sufficient information. Others noted that the document, while for the most part well-done, skimmed over some key environmental effects and dismissed others as nonexistent.

The document lacked credibility in other key areas. In particular, it failed to support a conclusion that the range was needed; assertions about effects on wildlife were not backed up by evidence; noise analyses in the study relied on inadequate methods; and it did not correctly analyze the effects on American Indians or historical and archaeological sites. Critics said the shortcomings called the validity of the entire document into question.

Though it did not include the restrictions he had repeatedly promised, Andrus lauded the impact statement and the cooperative efforts of state agencies and the Air Force. He reiterated his belief that the range was necessary to ensure the future of the Mountain Home base and assured that the range would be managed by the state, not the Air Force. The range would not use any live ordnance and would not allow supersonic flights lower than ten thousand feet above the ground. Fish and Game biologists had been consulted on the placement of target areas, Andrus said. And he told people the range had been modified in response to their comments and concerns that had "heightened our sensitivity to protecting the unique canyons and wildlife of the Owyhee desert."[9] The range would avoid sensitive wildlife habitat and river canyons, he said. His support of the proposal made it hard for critics to convince the public it was a bad idea. As Craig Gehrke noted, Andrus cast a long shadow. People liked him and they wanted to believe him.

But the plan outlined in the impact statement did not include any of the restrictions Fish and Game officials had requested. And the 166,000 acres of buffer around the 25,000 acres of bomb impact area were gone from the proposal. That made it look smaller, though it still would affect nearly 3 million acres of desert.

BLM officials said the impact statement did not comply with the Endangered Species or the Historic Preservation acts. Scrutiny brought by the range proposal had uncovered the archaeological richness of the north half of the range. More than 450 sites were found, some as much as five thousand years old, 15 of them eligible for the listing on the National Register of Historic Places and 112 more that might be eligible after further study.

The release of the draft impact statement brought focus to the Owyhee Canyonlands Coalition's efforts. The Air Force expected to make a final decision on the range by March 1994. Now there were public hearings to prepare for, presentations to arrange, and workshops and publicity campaigns to organize. Opponents had a target again.

Rick Johnson talked about plans for paid advertising and publicity campaigns. The Foundation for North American Wild Sheep, which had been working on its own, joined the coalition and members started showing up at the weekly meetings. The coalition organized workshops to help people prepare for the public hearings scheduled in January. The sheep foundation suggested a rally on the Statehouse steps during the hearings. Brian Goller outlined the necessary elements that should be included in the workshops.

Two BLM employees showed up at one meeting. They were evidence of the growing opposition among those who knew the Owyhee Canyonlands or were becoming familiar with the area as a result of the scrutiny brought by the state proposal. Though the BLM and other agencies were officially neutral, many agency employees increasingly were expressing their opposition as private citizens.

In a letter to Interior Secretary Bruce Babbitt in October, a group of range opponents likened the proposal to the Sagebrush Rebellion of the early 1980s. In a 1980 article, "They're Fixing to Steal Your Land," outdoor writer Ted Trueblood helped expose the plans of the so-called Sagebrush Rebellion to turn federal public lands over to state and private control for the private gain of miners, ranchers, and others. Like Trueblood, the leaders of fifteen groups opposed to the range warned Babbitt that "they're fixing to steal your land." They asked him to insist that all land withdrawals be brought before Congress for proper review—with the interest of all Americans considered.

"On one hand Governor Andrus claims the Air Force will close Mountain Home without a new training range," the letter said. "The Defense Department tells the (General Accounting Office) that existing facilities at Mountain Home are adequate and nearby existing ranges can be used. Then the Air Force states that existing facilities at Mountain Home are adequate now but in the future more will be needed, and that existing ranges are too far away. Then the Air Force tells a federal district judge that the (proposed) range isn't for the Air Force but rather the Idaho Air National Guard."[10]

The letter pleaded with Babbitt not to let Andrus and the Air Force steal the heart of the Owyhee desert.

Chapter 8 / Virtual Wildlife Refuge

THE BLUE SHADOW of the Owyhee uplands lies across the southern horizon, rising gradually to the west toward the Owyhee Mountains. A gravel road heads south, beyond the farmlands that stretch along the south side of the Snake River. The road winds up an escarpment and into a rugged lava landscape studded with sagebrush, old junipers, stands of mountain mahogany, and broken basalt rocks. Silence rings in one's ears. This high desert stretches across seemingly endless space to distant mountains. There are no fences or utility poles. But for the road and the ubiquitous bovines, man has left little mark here.

Vague tracks cross the rolling sagebrush with little to suggest the fantastic canyons that plunge deep into the landscape between the desert and the mountains. Out there somewhere the land drops unexpectedly into the thousand-foot canyons of the Owyhee River's East Fork and its tributaries. Sheer rock walls plummet to another world, where silver ribbons twist and turn between the ocher walls, where lush vegetation along the streams provides habitat for deer, elk, and other wild creatures. The rocky canyon rims shelter bighorn sheep, and thermal updrafts lift hawks and eagles on lazy circles. The peregrine falcon—making a slow comeback from the brink of extinction—has been sighted in these canyons. And some say the river offers one of the finest desert wilderness canoe experiences in the country. There is enough wild land here to accommodate a twenty-two-day canoe trip. The only other place in this country where this is possible is in the Grand Canyon. This is the Owyhee River canyon country—wild, unforgiving—whipped by freezing winds in winter and scorched by the sun in summer.

"The best way to see this country is from a canoe," said Phil Lansing, a river guide from Boise. "It's like driving around in an Ansel Adams photograph. This really is the outback, you know."[1]

Few people in Idaho know the canyons better than Lansing. Most didn't know of their existence until the state's range proposal focused attention on the area. But like Lansing, many of those who did know the area worked hard to spread the word about the natural beauty, wildlife, and other surprises to be found in the twisting canyons lurking unsuspected in this dusty green and gold sagebrush lava-rock desert. All this, they feared, was threatened by the proposed range. But the canyons and the lands above them held a few surprises even for wildlife biologists who knew the area well.

The bighorn sheep of those canyons had become the poster child for the effort to halt the proposed Air Force bombing range in the Owyhee Canyonlands. But raising sheep for hunters was not the only reason to protect the area. The sheep also held aesthetic value to those who camped, hiked, and floated the rivers. They were the most visible symbol of the effort to protect the fragile desert eco-system, but protecting the area meant protecting all the other species as well—from the hairy woodpecker to the sage grouse, from the mule deer to the elk.

In the late 1980s and early 1990s, Idaho Department of Fish and Game bi-ologists annually captured forty to sixty bighorn sheep—some years one hun-dred—from the herd in the Owyhee River canyons. When any herd reaches the limit that its habitat can support, animals become more susceptible to disease; thinning the herd helps maintain its vigor and keep the population stable.[2] The bighorns were transplanted to help establish herds in other suitable locations where they once thrived.

Ancient petroglyphs made by inhabitants of the Owyhee desert thousands of years ago often depict bighorn sheep. The journals of early nineteenth century trappers in the Rocky Mountain region tell of an abundance of the animals. They once ranged from the badlands of the northern plains to eastern Oregon and from New Mexico to Canada. Their range included most of the canyons of the Owyhee River drainage. As a ram can weigh up to 250 pounds, the sheep provided meat for early trappers, explorers, and settlers. And they were easy to hunt. In 1832, explorer Capt. Benjamin Bonneville reported that a few hunters could surround a flock and kill as many as they wanted. They found the meat better than the finest mutton.[3]

By the mid-nineteenth century, settlers began to stay in Idaho, and with them, their domestic livestock. Since then, unrestricted hunting during mining days, competition with livestock, and disease transmitted from domestic sheep all but wiped out the wild sheep. By 1900, most of them were gone from south-ern Idaho. The California bighorn subspecies (*Ovis canadensis californiana*), a relative of the Rocky Mountain bighorn sheep (*Ovis canadensis canadensis*), was gone. Fewer than one thousand Rocky Mountain bighorns were left in herds along the Salmon and Selway river drainages of central Idaho.

Reintroduction in southern Idaho began in the fall of 1963 when Idaho Department of Fish and Game trucks hauled metal corral sections and nineteen bighorns down a nearly forgotten wagon road through a dry side canyon to the flats along the Owyhee River's East Fork near the confluence with Deep Creek. Part of a gift of forty sheep, trapped near Williams Lake in British Columbia, they were left in the corral overnight to help them acclimate. But they kicked down the temporary corral. Later groups of sheep were simply released from trucks and trailers on the canyon rim above. They found their own way down—with far less difficulty than Fish and Game trucks.

In 1965, nine more sheep from Williams Lake were released in the same area, and in 1966 ten more. In 1967, twelve were released in Little Jacks Creek to the northeast. Fourteen more were released in nearby Big Jacks Creek in 1988,

Petroglyphs on the Owyhee Plateau. *Photo by Craig Gehrke.*

and twenty-four more in 1989. Subsequent observations showed the herds flourished in the canyons. In 1991, Lloyd Oldenburg, who managed the sheep program in Fish and Game's wildlife bureau, estimated the herds in southern Owyhee County at from twelve hundred to fifteen hundred. They made up twenty percent of the world's population of California bighorn sheep.

Today, these herds provide transplant animals for other former sheep habitats in the western United States, where, as in the Owyhee Canyonlands, white settlers and diseases long ago pushed the animals to extinction. Bighorn sheep from the Owyhee canyons have helped reestablish herds in the Jarbidge River and Bruneau River canyons and in Big Cottonwood Creek farther east in the South Hills, southeast of Twin Falls.

Bighorn sheep forage in open grassland and shrubland, where perennial bunchgrasses comprise most of their diet. They seldom stray more than half a mile from escape terrain or a mile from water. They generally do not migrate, and they favor areas with cliffs, steep slopes, and rock outcroppings where they can elude predators and find secure bedding places. Ewes give birth in the most rugged, isolated, and precipitous parts of their range with water and forage nearby. The ewes stay with their lambs in the area through the lambing season, which lasts from April through June.

The Foundation for North American Wild Sheep, based in Cody, Wyoming, and Idaho Fish and Game have invested heavily in the sheep herds of Owyhee County. A portion of the money for sheep transplants, for tracking the herds, and for habitat improvement comes from the sheep foundation. The foundation's Idaho chapter has raised nearly $500,000 by auctioning bighorn sheep hunting tags for tens of thousands of dollars each.

No one knows the long-term effects of the increased activity from the range. But research has shown that any disturbance that causes alertness and readiness for flight can cost the sheep energy, and at critical times of the year, that can be fatal. Fleeing from a disturbance also means the animal spends less time foraging. Repeated disturbances may drive the animals into poorer habitat, making them more susceptible to predators, or weaken them to the bacteria, viruses, and parasites already found in the herd.

Biologists predicted low-level overflights, sonic booms, and human intrusion from the proposed range operations would result in losses of fifty to eighty percent of the reestablished sheep herds in Owyhee County. The Air Force environmental impact statement's claim that the range would cause little long-term harm to the bighorn sheep, however, was based on the absence of definitive long-term studies.

The impact statement asserted that a cited study showed desert bighorn sheep adapted to the noise of low-flying jets over time. But the study's author, Paul R. Krausman, wildlife ecology professor at the University of Arizona, was quick to point out the limitations of his study. The study considered only the effects of jet overflight on sheep in a seven-hundred-acre enclosure on a mountain in Arizona. Krausman said that under similar circumstances he would expect the same responses in Idaho's California bighorn sheep. He acknowledged that heart rates "were significantly higher" during overflights. The sheep were startled and ran a short distance but settled down soon after the overflights.[4] But in Idaho the sheep live in canyons and the flats above the canyons, not in mountains as the sheep in Krausman's study did. Flights over Krausman's study area totaled approximately two thousand a year. The proposed flights over the Idaho sheep range numbered approximately ten thousand a year—and they would include frequent sonic booms.

Concerned about the accuracy of the impact statement, opponents of the range hired their own wildlife experts—experts who would be credible to both sides. Valerius Geist was head of environmental sciences at the University of Alberta in Calgary, Canada. He was recognized as one the world's foremost experts on bighorn sheep. Douglas Gladwin of Sterna Fuscata, Inc. of Fort Collins, Colorado, was a wildlife biologist and recognized expert on the effects of aircraft noise on wildlife. Gladwin and his partner, Alexie M. McKechnie, also a wildlife biologist, had studied the effects of low-level military flights on wildlife.

While Gladwin worked as a researcher for the U.S. Fish and Wildlife Service, he developed a way to predict and mitigate the effects of low-altitude aircraft on wildlife. He also had testified in federal courts as an expert witness on the effects of human disturbances on wildlife. Gladwin said conclusions reached in the environmental impact statement on the state's range proposal were contradicted by other studies and opinions not considered in the document. Since the impact statement neglected to cite or make use of studies that suggested potential adverse effects on wildlife from military overflights, he felt the Air Force was less than objective in disclosing the available scientific opinions.

"Considering the importance of the California bighorn sheep population in Owyhee County and the range of possible significant impacts to this population, little attention is given to this subject in the Air Force EIS. It is clear from a number of studies that bighorn sheep show both physiological and behavioral responses to aircraft overflights, suggesting that they are disturbed by the sight or sound of aircraft."[5]

In a separate critique, Geist also noted the shortage of data on the sheep and other animals and predicted that the range would drive bighorn sheep from their habitat in the Owyhee River canyons. The studies cited in the impact statement were not relevant to free-living bighorns. Penned sheep could not escape the overflight noise by moving or abandoning their home range. Mountain sheep are sensitive to loud, rumbling noise that in their native environment indicate avalanches or rock slides. Reaction to such noise may be an instinct, and if it is, the animals are not likely to habituate to it, Geist said. And California bighorn sheep appear to react more severely to disturbances than Rocky Mountain sheep. A study of California bighorns in the Fraser Canyon in British Columbia failed when the animals did not emerge from the canyon and resume normal behavior for weeks after being collared. The animals were eventually recaptured, the radio collars removed, and the study abandoned.

The Air Force impact statement assumed that because the sheep had been overflown in the past, they were habituated to overflights, Geist said. Because they lived under a training area did not mean they were not disturbed by military activity. If the habituation had been studied, it should be referenced, he noted. If not, the Air Force was wrong to assume the noise would not bother the sheep. Whether they habituate to repeated overflights was not certain, and what effects increased frequency of overflights would have over the long term was not known. "The Air Force, presumably has other air weapon ranges inhabited by mountain sheep. If so, why are there no studies submitted by the Air Force showing the fate of such populations?" Geist asked.[6]

Officials said Air Force ranges were among the most well-managed and protected areas in the country. But no one, including the Air Force, knows what happened to the two herds of desert bighorn sheep—a subspecies similar to the California bighorns—at the China Lake bombing range in California. They disappeared after the area was closed to the public. Some officials say the sheep died off when other animals overgrazed the area. Skeptics say poor management and military operations killed off the sheep.

Some of the best sheep habitat in California is in or near other military operations areas. Operations there apparently have not harmed desert bighorn sheep. But they do not face the severe winters that the Idaho sheep do. And winter is particularly stressful for the animals. If the sheep have stored enough energy reserves to last through the winter, they survive. If they burn up the energy before spring arrives, they die—or they are weakened and succumb to disease or predators.

From 1988 to 1993, flying the same route each year, Fish and Game counted between six hundred and seven hundred bighorn sheep annually in Owyhee County. In 1994, the count dropped to 346. Harsh weather, a dry summer in 1992 followed by deep snow, may have contributed, biologists speculated; blood tests of captured sheep showed no evidence of disease. But critics of the proposed range noted that the sheep numbers were still up in 1993 following the winter and that military overflights increased and sonic booms began later in 1993. There had been severe winters and dry summers in the years before 1992. Some thought the increased activity and repeated disturbance were driving the sheep into poor habitat. Before the increased activity began, biologists had predicted that jet noise and sonic booms would cause some sheep to abandon their traditional habitat. The Air Force maintained there was no direct evidence of any relationship between overflights and bighorn sheep numbers.

In June 1993, an aerial survey of the high desert of Owyhee County turned up a big surprise. The number of pronghorn antelope north of the Owyhee River's East Fork and within the proposed bombing range was estimated at 1,708—far higher than anyone had thought. The area had the highest density and highest productivity of pronghorn in Idaho.

Pronghorn fawn survival is higher in low-sage, wet spring habitat than in surrounding areas. This favorable habitat is limited and does not exist in significant amounts outside the area of the proposed range, Fish and Game biologists said. They predicted the range would result in herd losses from twenty-five to seventy-five percent. Some of the negative effects of the range would be hard, if not impossible to mitigate. Forage lost to construction, fire, and fencing could be partially replaced outside the range area, but forage was not the limiting factor for pronghorns. It would be impossible to replace the favorable fawning habitat. If such habitat existed in the vicinity, the antelope probably already were using it. Increasing forage elsewhere would not make up for predator losses if does were forced to give birth in areas where the fawns would be more vulnerable.

The fleet pronghorn is found throughout southern Idaho. Its greatest numbers are in the Big Lost and Little Lost river valleys, Lemhi and Pahsimeroi ranges, and in the Owyhee highlands. They prefer open sagebrush and grassland plains, and they migrate from the cooler high plateaus to warmer valleys in winter. Their keen eyesight and speed—anywhere from thirty-two to fifty miles per hour—are their main defenses against predators. Pronghorn eat a wide variety of shrubs and herbs—competing with domestic sheep—but mostly they eat shrubs, some sagebrush, and some rabbitbrush. They survive on as little as one to three quarts of water per day.

Pronghorn mothers give birth in the spring. They select birth sites that protect the fawns, which cannot travel much until they are about two weeks old. They spend most of that time bedded down in hiding. The does move off up to a mile away to feed and to rest. They will visit their fawns several times a day to nurse. During this time fawns are highly vulnerable to golden eagles, coyotes, and bobcats. But in the low-sage flats away from canyons, wind currents are not favorable for soaring. Does aggressively protect their fawns, and in the low-sage flats they can see predators from farther away. This combination of low-sage flats away from hills and canyons is rare outside the proposed range area. Though antelope have flourished with past military activity, it is uncertain how they would fare with the increased concentration of overflights and on-the-ground human intrusion. They are sensitive to aerial predators, and during fawning does often chase golden eagles and large hawks. They have been known to do the same to small, low-flying aircraft.

The high plateau of the Owyhee Canyonlands also contains one of the greatest number of historical undisturbed sage grouse leks in southern Idaho. And those leks, or strutting grounds, may hold the key to the continued survival of this increasingly rare desert bird. Sage grouse populations have been declining in southern Idaho and the West, and talk of considering the bird for endangered species status already had started.

The size of a small turkey, sage grouse are the country's largest grouse, weighing up to seven pounds. They have no gizzard like other birds, and so they rely on soft plant parts. They favor certain strains of sagebrush, sometimes even individual plants. In spring, insects form a key part of the young chicks' diet. They also depend on grasses and small plants for nesting and raising their young. In winter, sagebrush provides food and shelter from the cold. In areas with adequate sagebrush, sage grouse are seldom affected by severe weather. Traditional nesting areas, leks, and winter habitat all are found where the birds can find their preferred food.

Activity at the lek begins at first light. Male sage grouse puff up air sacs on their throats, raise their spiked tails, and bob and weave while making strange bubbling, popping noises, hoping to attract willing females. The males continue to gather at the lek for about two weeks after the substance of this mating ritual is finished and the females have retreated to nest. Nests are located nearby in tall, dense stands of sagebrush. A hen lines a slight depression in the ground under a sagebrush with a little grass and a few feathers. About ten days after mating, she lays six to nine olive, yellowish eggs evenly dotted with browns. Chicks begin hatching as early as mid-May and into early June.

Some research shows that sage grouse abandon leks for a day or more if disturbed by a loud noise. After repeated disturbances they may abandon the lek completely. And nesting females may abandon their nest after a single disturbance. The Air Force impact statement suggested that sage grouse could move to another area if displaced from leks and nesting areas. Long-term research, using leg bands and radios to track the birds' movements, shows sage grouse move around considerably. But they are not migratory—they like to come back to the same leks year after year. If the leks are disturbed, the birds disperse. Biologists do not know what happens then.

Some of the biggest surprises on the high and otherwise dry Owyhee Plateau are wetlands—marshy areas near springs, along the streams, and in the small seasonal lakes known as playas. Some are man-made reservoirs that trap seasonal moisture for livestock. Others are natural low spots that catch melting snow and rainwater—some years they dry up early, other years they hold water through the summer.

Though written off as only stopovers for migrating birds, it is exactly such areas along migration routes that allow migratory birds to survive. The local disappearance of a few water holes might seem insignificant—birds are there only a few days or weeks a year—but it is wrong to assume that less-heavily used areas are unimportant.[7] The decline in the number of shorebirds in recent years is in large part the result of the piecemeal destruction and disappearance of just such wetlands. Birds use the wet areas for staging and to rest for the next leg of their journeys. They spend much of the time feeding to gain energy for the coming strains of travel to their breeding or wintering grounds. The wetlands in the Owyhee Canyonlands provide habitat for several migrating bird species, including northern pintails, American widgeons, green winged teals, long-billed dowitchers, American avocets, and white-faced ibis. For untold centuries, these birds have depended on the few scattered water holes in this desert, where they have found sustenance and rest on their peregrinations.

The key to this abundance of wildlife in the high desert of Owyhee County is in part the diversity of native plants. The area is one of the few in the state where large expanses rich in native plant life have not been destroyed by fire and overgrazing as in more accessible parts of the West, those lower in elevation. The high desert supports about seven hundred species of grasses, small plants, shrubs, and trees. Only a few are rare.

But Air Force operations and increased human activity would increase the likelihood of fire and leave disturbed soil—both an open invitation to invasion by exotic weeds. And the dry Owyhee Plateau is susceptible to dramatic change by fire. Though a natural part of the desert ecosystem, fires give invading weeds a place to start—especially cheatgrass, introduced in the American West from the steppes of central Asia more than one hundred years ago. Fire favors cheatgrass, which uses fall moisture to germinate and turns green and matures early in the following year. The dry grass burns quickly and easily, leaving burned areas susceptible to the further spread of cheatgrass, which in turn leads to more frequent fires. Perennial grasses and many other native plants cannot endure frequent fires, and eventually diverse stands of native grasses and sagebrush are reduced to little more than cheatgrass and tumble mustard—and some noxious weeds.[8]

One January day in 1993, while biologists were studying the wildlife carrying capacity along Deep Creek and the East Fork of the Owyhee River for the Air Force's environmental impact statement, one of them spotted a hairy woodpecker. The robin-sized black and white bird was pecking away at a snag when a sonic boom shook the area. The bird was thrown backward off the tree trunk. It regained its composure and flew off through the woods, screaming a distress call. The woodpecker continued making noise long after it disappeared from sight. The biologist in his report noted that, apparently, sonic booms temporarily affect foraging in this species.

It is impossible to believe that sonic booms and jet overflight noise would not affect other species in a similar manner. Though the animals may be holding their own—or even thriving—no one knows how much additional activity they can endure before their numbers begin to crash. The state's proposal focused attention on the Owyhee Plateau and showed how little was known about the area and its wildlife, and how little was known about the effects of aircraft noise on wildlife. That lack of knowledge, however, does not mean low-flying and supersonic jets have little or no effect as some have claimed. That assumption, biologists note—as did the woodpecker—may not hold true.

Chapter 9 / Mounting Challenges

T HE FIRST PUBLIC HEARING on the draft environmental impact statement for the range proposal was set for January 12, 1994, in Boise. In anticipation, opponents of the proposal had mounted a publicity campaign. They rented seven billboards in Boise, two in Nampa, and one in Twin Falls—at a cost of about ten thousand dollars. The billboards carried a stylized design of the Owyhee River Canyon with the silhouettes of two low-flying fighter jets on the right, two bighorn sheep on the left, and a simple message: "Owyhee Canyonlands: No Bombing Range!"

Owyhee Canyonlands Coalition members organized a public workshop on January 8 to help people prepare for the hearings. Bob DiGrazia of the Foundation for North American Wild Sheep discussed the likely effects of the range on recreation and hunting while Herb Meyr covered wildlife and presented his slide show. Phil Lansing and Wendy Wilson of Idaho Rivers United talked about white-water recreation opportunities in the rivers of the Owyhee Canyonlands, and Brian Goller explained how to testify. The Snake River Alliance organized a presentation in Boise by the Shoshone-Paiutes. More than two hundred people showed up to hear about concerns for tribal lands and how the range would affect religious sites and practices.

Much of the money for the January publicity campaign was donated by private citizens and groups, including Theresa Heinz, the widow of the late Sen. John Heinz from Pennsylvania (of H.J. Heinz ketchup fame), Christopher Hormel (of Hormel meat fame), and the Aldo Leopold Society of San Francisco. Donations paid for full-page ads in national newspapers and television air-time for ads that featured actors Scott Glenn and Mariel Hemingway and Tribal Chair Lindsey Manning of the Duck Valley Shoshone-Paiutes speaking against the proposal. Advertising director Joseph Hanright donated his time to help put the ads together. The Sheep Foundation came up with sixty-five hundred dollars for Bob McEnaney of the Snake River Alliance to organize a series of radio ads featuring Hollywood stars Pamela Sue Martin, Adam West, and Scott Glenn. Full-page ads were taken out in the *Idaho Statesman* in Boise and the *Boise Weekly*.

The campaign focused public attention on the resources threatened by a bombing range and training complex that many felt was not needed—even Air Force and Defense Department officials could not agree on the need. Coalition members lined up critics, attacked the adequacy of the impact statement, and charged that the state's proposal bypassed congressional scrutiny. Some members wrote directly to high government officials. Together, the coalition tried to raise public awareness and get people to attend the public hearings. Apparently they succeeded—172 people signed up to speak at the hearing in Boise, forcing the Air Force to schedule an additional day.

On January 12, the Air Force set up its booths in one of the ballrooms at Boise State University's student union. The displays, maps, graphics, pictures, and textboxes again made the event look more like a trade show than a public hearing. Air Force officials talked about the need for the range, noting that it had been cut down from 1.5 million acres of the ill-fated Saylor Creek proposal to only 25,000 acres.

Across the hall, the Owyhee Canyonlands Coalition had rented space to set up what members called their "Range Reality Room." Maps, pictures, and handouts presented the coalition's perspective. One map showed the range as approximately 25,000 acres of state lands surrounded by buffer zones on BLM land amounting to 96,000 acres north of the river and 70,000 acres south of the river. Another map showed the proposed ranges, the existing Saylor Creek range, and the thirty mobile radar units that together would create an electronic combat range spread over 3 million acres.

Meanwhile, in U.S. District Court in Boise, a lawyer representing the Air Force testified that the range was not a necessity.

Then on a Saturday morning late in January, not long after the first public hearings, more than two hundred volunteers crowded into the Boise YMCA. They received goodies—coffee, fruit, and drinks—pep talks about the importance of the issue, and a little traveling music from three local musicians. They were out the door within forty-five minutes and spread out throughout the city, where they distributed more than twenty-eight thousand doorhangers. The effort made the evening news, adding to the effectiveness of the campaign. The doorhangers also were distributed in seven other Idaho cities. The greatest challenge had been to get enough people involved and still keep the effort quiet enough to take advantage of the surprise.

The doorhangers carried the same design and the same direct message as the billboards. On the back were two tear-off postcards, one to Idaho state Congressman Larry LaRocco—"Owyhee Canyonlands is of national significance. The U.S. Congress should provide permanent protection for this prime wild desert area."—the other to Interior Secretary Bruce Babbitt—"Owyhee Canyonlands deserves the highest protection possible by the BLM. These lands must not be traded to the state of Idaho for a bombing range."

Meanwhile, the coalition had hired former Pentagon economist William Weida, a specialist in defense economics, to analyze the proposal. Weida said the impact statement provided no evidence that training in Idaho would be more efficient than training anywhere else. The criteria that required a range to be within 150 nautical miles were chosen without evaluating the costs and benefits of a range father away. He also pointed out that Owyhee County would incur all the costs in the form of reduced recreation, noise, dust, fires, and chaff, while Elmore and Ada counties would reap the economic benefits.

Noise and effects on lifestyle would reduce Owyhee County property values—a total assessed value of about $250 million—by as much as $750,000, Weida said. The land swap proposed by the state would result in further erosion of the county tax base. The proposed range also would eliminate about $127,000 worth of grazing permits. Balanced against the benefits of "national security," Weida estimated the proposed range would cost the county about $330,000— which translated to about $40 per person. And Owyhee County could least afford the cost, being one of the poorest counties in Idaho—twenty-nine percent of the population lived under the poverty level.[1]

Weida joined others in attacking the Air Force's noise analysis, which relied heavily on a fifteen-year-old study that was based on reactions to noise in urban settings, mixing aircraft, street, and rail noise. The study covered people in buildings in cities and had little bearing on the annoyance of jet noise in a wilderness setting. One noise expert noted that the analysis in the impact statement was done by unqualified staff.[2] That alone was enough to raise questions about the conclusions, predictions, and estimated effects of aircraft noise. No baseline of existing conditions was established. Some noise figures presented in the document were not measured but calculated using a computer program that has since been shown to underestimate noise predictions by more than six times. Overall annoyance may have been nearly twelve times greater than the model predicted, according to a preliminary Air Force study on low-level flights.

The true effects of aircraft operations were hidden by averaging noise over twenty-four hours, depressing large numbers. The impact statement presented numbers only for individual aircraft on a single, brief overflight, not the repeated overflights of multiple aircraft in the exercises proposed by the Air Force. The Idaho impact statement listed the noise of an F-16 flying slow and level as 97 decibels at one thousand feet. But an impact statement for the Pennsylvania Air National Guard listed the noise of an F-16 during operations at 111.6 decibels. Reality at times in some places under the proposed operations would generate noise exceeding 120 decibels, enough to damage human hearing.

The impact statement also quoted a 1992 Forest Service study, "Potential Impacts of Aircraft Overflights of National Forest System Wildernesses," to back an assertion that aircraft noise does not significantly affect a person's wilderness experience, though the study suggested that people in remote wilderness areas are ten times more sensitive to noise intrusion than those in an urban area and its author said it could not be used to predict the effect or annoyance of low-flying military jets.[3] Only two percent of the aircraft observed in the study were military jets, and only a few of those were low-flying. It is impossible to generalize from those numbers, he said. The Air Force may have assumed that meant nobody noticed the military jets and thus they would not affect wilderness recreation values. But the Air Force missed one important observation. Those few who were exposed to low-flying jets reported them as by far the most annoying aircraft.

The inaccuracy of the noise analysis Air Force planners had used in the impact statement on the state's range proposal and other proposals at the time, eventually was acknowledged by the Air Force in its own "1996–97 Environmental, Safety and Occupational Health Research, Development and Acquisition Strategic Plan." That document also noted the analysis method's vulnerability to challenge under the National Environmental Policy Act. The strategic plan's purpose was to outline environmental health and safety needs and how those needs could be met with research at Department of Defense laboratories. One of those needs was "legally defensible" noise analysis methods to establish, maintain, or change military airspace and to predict the annoyance of sporadic sonic booms over largely dispersed populations. The current methods of analyzing annoyance had not been shown to be applicable in rural areas of practical interest to the Air Force, the document said. Noise analysis methods were developed for urban areas, not wilderness or recreational areas or rural areas with dispersed populations. And computer programs for predicting noise effects relied on data gathered and entered by humans, which made the outcome vulnerable to human error and therefore to legal challenges.

The Air Force should have acknowledged these limitations in its impact statement. Perhaps Air Force leaders were reluctant to acknowledge that near an exercise, low-flying jets would make enough noise to damage a person's hearing. And at the time, the Air Force study of the issue still was in its early stages.

In early 1994, some coalition members got busy with their typewriters, including Fred Christensen, president of the Idaho Wildlife Federation. He wrote well argued letters to key elected officials, including Vice President Al Gore; Kathleen McGinty, head of the White House Office on Environmental Policy; Interior Secretary Bruce Babbitt; Defense Secretary William Perry; and Undersecretary Sherri Goodman. But perhaps Christensen's most telling letter was to Air Force Secretary Sheila Widnall. He took her to task for calling the proposed range critical, crucial, and necessary.

"Based on your department's own statements, it is clear that the proposed range is none of those things and that your statements are, at best, misinformed, and, at worst, part of the Air Force's long campaign of half-truths, misinformation, and dissembling about the need for the range and its impacts."[4]

Air Force officials had repeatedly said it did not need the additional Idaho range to train the composite wing. A Defense Department official had said the ranges in Utah and Nevada would provide more realistic training by requiring tanker support for the training that could not be done in Idaho. Since the end of the Cold War and the collapse of the Soviet Union, Pentagon officials had not been able to show a need for additional training facilities. In fact, one official acknowledged that a careful training needs assessment most likely would show a need for less training space. But the Air Force had not done such an assessment nor had the Department of Defense.

In the early 1980s, the evaporation of the primary military threat to the United States and an ever-growing budget deficit led to a substantial reduction in military forces nationwide. The Air Force withdrew from twenty-eight overseas bases and closed nineteen U.S. bases—reducing by about one-third its number of aircraft, personnel, and bases. Yet at the same time, Air Force leaders were planning expansions in Idaho—up to sixty million dollars in base improvements at Mountain Home—and in other Western states. Air Force leaders planned to cut thirty-six fighter wings down to twenty-six, a reduction of 720 fighter jets. But rather than cut numbers across the board, military leaders wanted to keep fewer units intact, capable of fighting and winning a war against greater numbers. The composite wing would be one of those units, and Mountain Home would be one of ever fewer Air Force bases as the military tightened its belt.

But even within the military, officials continued to disagree about the need for additional training facilities in Idaho. Many officials had acknowledged what the Air Force wanted in Idaho was the wide-open airspace that stretched into eastern Oregon and northern Nevada. A new or expanded training range was merely a way to secure that airspace and the air base through future rounds of base closings.

While other officials said the new range in Idaho was not a necessity, in late 1992 composite wing commander at Mountain Home Brig. Gen. William S. Hinton Jr. had said that failure to expand training facilities would harm "mission readiness." He refused to explain how. Further uncertainty about the need for the range arose from a May 1993 General Accounting Office report that questioned the wisdom of establishing the untested composite wing. In its response to the GAO report, the Defense Department asserted that existing range facilities in Idaho and neighboring states would accommodate the wing. The existing airspace in Idaho and the Saylor Creek range would provide about eighty percent of the wing's training. For complex composite force training that required ground targets, the wing would fly to ranges in Utah and Nevada, which is what Air Force officials had told the Base Closure Commission—that those out-of-state ranges would accommodate the training that could not be done in Idaho.

But Air Force officials were telling people in Idaho that it was too far for the wing to fly to those ranges every day, that it would be inefficient, though by the Air Force's own admission, pilots would fly to those ranges only a few times a month—for four to five hours of training a month—not every day. In addition, the proposed Idaho range would accommodate only about half of the training now done at those ranges. Pilots would still have to fly out of state to practice with live ammo. And some officers were proud to point out that even without the state's range, the wing was the best-trained unit in the Air Force. Training already had prepared the wing to go anywhere in the world on a moment's notice, they said. It was obvious to most that the wing did not need the range to train.

One of the reasons flying out-of-state would be inefficient, Air Force officials said, was that the planes would have to link up with air tankers and refuel under way. But that is reality, Herb Meyr said, and the Air Force continually talked about wanting realistic training. The long flights help pilots learn how to manage the cockpit—where to stow the maps and charts, "piddle-packs," water, snacks, and other items needed on long flights and in real combat. Pilots must know how to hit their targets at the correct time, elude enemy defenses, and find their way back to the tanker and home if they are to survive in the real world.[5]

The Air Force cannot fight a war without tankers, which extend the effective range of small fighter-bombers. The F-16, for example, cannot fly a mission without refueling, Meyr said. During the Persian Gulf War, U.S. fighter pilots would take off, meet up with a tanker for fuel, fly to the target, and drop their bombs or fire missiles. On the way back they met the tanker and sipped a little more fuel to get home. Learning fuel conservation in hostile areas and refueling over unfamiliar terrain is a basic requirement. "You definitely do not want to run out of gas over enemy territory," Meyr said. And pilots had to practice to become proficient at refueling in the air—even in bad weather, at night, or over unfamiliar terrain.

Aerial refueling already was an important part of the required training in Idaho. Pilots practiced the art of sipping fuel at twenty-six thousand feet and 350 mph on a regular basis—even when they did not fly to distant ranges.

One day in the summer of 1993, with a line of thunderheads threatening to the south, an Air Force KC-135 waited for takeoff clearance from the control tower at Mountain Home Air Force Base. Pilot Capt. Ken Shaffer worried that the fighters, with which he was to rendezvous high over the desert, would be unable to fly to take on the fuel he was carrying. And that would leave him dodging thunderheads for eight hours while he burned enough fuel to land the tanker safely. But then a pair of sleek F-15s screamed down the runway, taking off in tandem just a few feet apart. Moments later two more fighters followed. Then it was the tanker's turn. The late-1950s Boeing 707 had newly replaced engines and fuel tanks in its belly where the commercial airline version would have carried baggage and freight. Shaffer opened the throttles. The great beast shook, and the day's mission was under way.

From his perch deep in the belly of this great gas station in the sky, Sgt. Troy Doane, a twenty-five-year-old Texan, operated the boom lying on his stomach in the "boom pod" under the tail of the plane. Midair refueling, a delicate operation during daylight, becomes even trickier at night, when lights alone indicate where the thirsty jet fighter might be—at times the planes use no lights at all. But refueling at night in rough weather, when both fighter and tanker buck and toss, is "stressful," Doane said. He had hit the padded ceiling a few inches above his back more than once.

Once the pilot is lined up beneath the tanker's belly, it was up to Doane to thread the fuel receptacle without hitting the aircraft or its radio antennae and "get him his gas as quickly as possible." Inside the twenty-eight-foot boom, a four-inch diameter extension increases Doane's reach to forty-six feet. He guides it with a joystick, and hydraulic latches lock it to the fighter while fuel is transferred. It takes only a few minutes to fill the fighter's tanks. During the three-and-a-half-hour exercise that afternoon, the tanker pumped ninety-one thousand pounds of jet fuel into fifteen fighters from the Mountain Home base while flying a racetrack oval seventy miles long and twenty miles wide over northwestern Nevada.

After refueling, the fighters returned to Idaho for the day's training exercise. But as Meyr pointed out, flying to northern Nevada to rendezvous with the tanker and then assembling a strike force for an exercise over an Idaho range was about the same as flying to the Utah Test and Training Range in northern Utah, about twenty minutes in the air from the Mountain Home base. Increased training in Idaho—if it were to be realistic—would be no more efficient than what the pilots already were doing, Meyr said.[6]

In addition to questions of need, a frequent comment during the January 1994 public hearings on the impact statement was that the state's proposal for expanding Air Force training facilities should have thorough congressional scrutiny. Critics noted that had the Air Force proposed establishing the range itself, the Engle Act of 1958 would have required such scrutiny. That act was the result of earlier unrestrained military expansion.

Between January 1955 and June 1956, the military had asked for more than fourteen million acres of public land. Those expansion proposals had caught the attention of Congress. Executive orders in 1942, 1943, and 1952 had delegated the implied withdrawal authority of the executive branch to the secretary of the interior. The Interior Department simply approved the military applications for withdrawals as they came in. Much of the land at the time was considered desert wasteland. But hunters complained to their congressional representatives when the withdrawals began to impinge on their favorite hunting areas.[7]

The U.S. Constitution grants Congress the power "to dispose of and make all needful rules and regulations respecting the Territory or other property belonging to the United States." Congressman Clair Engle of California introduced legislation, passed in February 1958, that returned that power to Congress. The Act required congressional approval for the "withdrawal, reservation, or restriction of more than 5,000 acres in the aggregate" of public land by the Defense Department, and it included the authority to set terms and conditions on the withdrawal and to require public hearings. That slowed the land grab, but it did not stop the military's appetite for land or airspace.

In 1994, state officials said the range they were proposing would be a state-operated Air National Guard range, and the Air Force would merely be a user. They said the state would retain ownership and control of the range, and they had no intention of circumventing the Engle Act. The fact that the swap would be more timely than a federal land withdrawal was incidental. But Congress had some doubts. A House subcommittee report pointed out that the state was acquiring land for the Air Guard, which had said it did not need a new range, and the range was being designed for the Air Force's newly established composite wing. If the Air Force tried to build the range itself, it would have had to get the approval of Congress. The subcommittee asked the secretary of defense to explain why this approach did not circumvent the authority of Congress under the Engle Act.

No one would argue that a new or expanded range would not improve training in Idaho. But the Air Force had not convinced anyone that training would suffer without the range. And in the absence of a clearly demonstrated need, the apparent attempt to avoid congressional scrutiny—whether intentional or not—raised additional suspicions in Idaho and in Washington, D.C.

Chapter 10 / Undercurrents

I N JANUARY 1994, a rumor circulated through the ranks at Fish and Game that Governor Andrus had told department director Jerry Conley that anyone who testified against the range at the upcoming hearings would be fired. A department memo went out a few days later to clarify that if employees chose to testify, they were not to identify themselves as department employees or claim to represent the department. The policy had been on the books a long time, though it had not been enforced during earlier Air Force hearings. Andrus knew that criticism of the proposal by professional wildlife biologists gave opponents credibility. But he denied having tried to muzzle Fish and Game. His intent had been to make it clear that the state officially endorsed a state-owned range, and he did not want anyone representing a state agency to say otherwise. No one was fired.

The incident nearly silenced Fish and Game staff, but cracks in the support that Andrus counted on were widening in early 1994. Fish and Game commissioners withdrew their support for part of the proposal. Bureau of Land Management head Jim Baca's row with Andrus extended those cracks to the White House, where support also was divided and subject to political influence. Baca's demise brought uncertainty for range opponents. To those in Idaho, it was unclear with Baca gone who would make the decision on the range proposal. Nor was it clear what was holding up the release of the environmental impact statement. But one thing became clear: the Department of Interior still would have to resolve the opposition of the Shoshone-Paiute tribes before the range could go forward.

During the first day of the Fish and Game Commission's two-day quarterly meeting in late January 1994, Fish and Game biologists presented a report that showed a surprisingly high number of pronghorn antelope in the north half of the proposed range. The report predicted the range would result in a long-term decline in the pronghorn population. Fish and Game staff also told the commissioners that basic restrictions, upon which the commissioners' support had hung, were not included in the impact statement.

During the meeting, the commissioners also heard concerns of the Owyhee Canyonlands Coalition, which had asked to be included on the meeting agenda. The coalition wanted to show the commissioners that many people opposed the range, hoping to make it more likely that the commissioners would resist pressure from Andrus. Approximately thirty people showed up.

"To some folks this is just useless wasteland, and they'd just as soon let the military drop bombs on it," Idaho Wildlife Federation Director Dick Juengling told the commissioners. "But the hundreds of people who go down there each year find it to be a place of unparalleled solitude, unparalleled enjoyment, and unparalleled inspiration."[1]

The commissioners had planned to make a decision about the range proposal on the second day of the meeting. But one commissioner was out of the state and could not be reached until late Friday. By that time, another commissioner had left to fly home. Instead, the commissioners put off discussing the public comments and the report by Fish and Game officials until the following Wednesday, February 2, in a telephone conference call. They also had wanted Andrus involved, but he was going to Washington, D.C., for the annual winter meeting of the National Governors' Association.

The commissioners went ahead without him. In a five-to-one vote, while Andrus was on the plane to Washington, the Fish and Game Commission pulled its support for the range proposal. Their position was presented as a cover letter to a summary of the department's comments on the draft environmental impact statement. The commissioners criticized the document as incomplete. They cited a lack of adequate consideration for the protection of "unique biological resources," including bighorn sheep, pronghorn antelope, and sage grouse. Based on the review by the department's wildlife biologists and the public testimony, "the Commission can no longer support the inclusion of the north training range as part of the preferred alternative."[2]

Norm Guth was the only commissioner to support the entire proposal. The rest of the commissioners might have pulled their support for the south half of the range as well, but problems could not be identified with the little information available. The commissioners, however, had clearly stated they would be willing to review other range proposals in locations less sensitive to wildlife.

The shortcomings of the impact statement were like a wedge driven between Andrus and Fish and Game, sworn to manage the state's wildlife. Fish and Game's comments echoed many of the concerns expressed earlier about the preliminary document. But this time Fish and Game had all the chapters. While true that the commission had supported the split-range proposal, further study had shown the proposal would potentially be more harmful to wildlife than first thought, that much of the wildlife habitat lost to the range could not be replaced anywhere. The commissioners felt a moral obligation consistent with their oath of office, and at risk of their positions, to protect the wildlife.

Fish and Game wanted two major issues resolved. First, many elements of basic mitigation were not identified in the impact statement or any other legally binding manner. "Assurances that these elements would be contained in the (draft impact statement) were replaced with assurances that they would be contained in the Range Operations Plan; to date they appear in neither."[3] Second, the document did not clearly identify who would be legally and financially responsible for mitigation and compensation as required by state policy. Mitigating adverse effects on wildlife would require a long-term commitment, and compensation for certain unavoidable effects could be costly. Department officials were concerned about the potential reductions in wildlife and plant populations, the loss of wetlands essential to migratory birds, and degradation of recreation opportunities in a nearly roadless, remote wild land within a day's drive of Idaho's most heavily populated counties.

Governor Andrus' liaison Dave Jett said most of the commissioners' concerns were addressed in the impact statement. Richard Meiers and other commissioners thought they were not and that the impact statement did not commit the Air Force to any action.

Andrus felt betrayed. The commissioners had prejudged the impact statement before Fish and Game's comments could be incorporated, he said. Andrus, himself an avid hunter, blasted the commissioners for rejecting the range proposal out of concern for wildlife while "they hand out a number of permits so wildlife can be killed."[4] He lashed out at Fish and Game Director Jerry Conley. The proposed bombing range was very important to the state economy, and Fish and Game should have considered economics when evaluating the impacts of the proposal on wildlife, Andrus said.

The province of Fish and Game is wildlife, not economics, Conley countered. The commissioners said they were unwilling to risk wildlife as long as reasonable alternatives existed. Their decision was a real blow to the credibility of the range proposal. National conservation groups could now say that even the Idaho Fish and Game Commission was against the proposal.

Mountain Home area lawmakers of both parties were outraged. State Reps. Robbi King of Glenns Ferry and Frances Field of Grand View and Sen. Claire Wetherell of Mountain Home charged that the commission was unethical in pulling its support, and they called for Conley's resignation. They said they would support Andrus if he were to fire the commissioners as well. King also complained that the commissioners should have let the public know they were wavering on the issue, thus encouraging more people to speak out in favor of the range during the commission's meeting. State Rep. Bruce Newcomb of Burley, not to be outdone, proposed cutting the Fish and Game budget by seventy thousand dollars, in part because he was upset that the commissioners had waited so long to express their opposition to the range proposal.

The day after the Fish and Game commissioners pulled their support, BLM head Jim Baca was fired. Many in Idaho thought it was over his row with Andrus. Baca, who later became mayor of Albuquerque, did not blame Andrus for his demise. Others in the administration wanted him out. He was offered a job as a deputy assistant secretary of the interior. The offer, couched as an ultimatum, was presented by Babbitt's chief of staff Tom Collier. Baca refused it and was fired on February 3. He said Collier had lobbied the Western governors to demand the White House fire him.[5] Baca had also angered other Western lawmakers with his unpopular stance on federal land grazing and mining. White House officials, sensitive to the views of other Western Democrats who in 1992 had endorsed Clinton, the first Democrat in a generation to carry many Western states, tried to distance themselves from the affair. They said the impetus for firing Baca came from Babbitt's office, not the White House.

Some thought Baca had showed the strength of will and dedication to public service required to do what very few public officials have done—demand due process and public involvement. He had struck a raw nerve when he pointed to what had been a political rather than scientific process. He was asking good, tough questions about the bombing range, said Craig Gehrke of The Wilderness Society. Many had seen Baca as their best hope to defeat the bombing range proposal and to create a National Conservation Area designation for the Owyhee Canyonlands, an area twice the size of Yellowstone National Park. With Baca's demise, many saw the hope of such protection fading.

Regardless of reasons, Baca's firing made him a martyr to the cause of the range opponents, and it catapulted the bombing range into the national spotlight in Washington, D.C. The issue was suddenly relevant inside the Beltway. But many thought the fight was lost. They were afraid Babbitt now would go along with the proposal. Though the decision on the exchange of thousands of acres of BLM land, necessary for the state's proposed range, normally would have been made by Baca as head of BLM, Babbitt assured Andrus that he would make the decision himself. And some skeptics were certain that Andrus had agreed to support Babbitt's mining and grazing reforms on public lands in the West in return for Babbitt's support on the bombing range.

Further fueling suspicions in March 1994, Tom Collier arrived in Boise unannounced to meet with Andrus and state BLM Director Delmar Vail. Collier had been arranging meetings with key players on the range issue, apparently working on a deal. Some thought the Interior Department was ready to give in to Andrus. Collier contacted the Washington office of The Wilderness Society, proposing an Idaho BLM wilderness bill covering at least the Owyhee Canyonlands in exchange for a promise not to sue if the Interior Department approved the land swap for the range. The national office contacted Craig Gehrke in Boise. The answer was no. The deal amounted to federal appropriation of land without due process, Gehrke said.

Collier tried to convince Gehrke that there was no other way the environmentalists would get a wilderness bill. Gehrke in turn pointed out that they had not had a forest wilderness bill in about twenty years, so why rush a BLM bill that no one supported but the Air Force? The offer amounted to trading wilderness for a bombing range. But without airspace restrictions protecting that wilderness, it was an empty promise, Gehrke said. What good was a wilderness beneath a war zone? Collier had not yet tried to contact other members of the

Owyhee Canyonlands Coalition. Perhaps he thought The Wilderness Society had the most legal and lobbying clout and influence over other coalition members. Perhaps he thought the environmentalists would be unable to resist a sure-win wilderness bill. But others in the coalition would be harder to convince. And if the range were approved, Congress still would have to consider the wilderness study areas affected by the range. All of them could be considered for wilderness designation, but a wilderness bill that included only some areas would drop other areas from further consideration. That would make a deal hard to sell.

At the same time, Collier was working on a letter approving the land exchange. The letter was to go to the White House in mid-April, with Collier's signature—apparently bypassing Babbitt. The letter set conditions on the exchange: the Air Force would have to resolve issues involving threatened and endangered species, archaeological values, the loss of wetlands, mitigation, and the designation of wilderness study areas. The National Environmental Policy Act, however, required the release of the final impact statement and a favorable record of decision before the range proposal could go forward. But apparently no one at Interior wanted to push the letter approving the range except for Collier.

Then in early April, word leaked out that the mission at Boise's Gowen Field had been changed from F-4 Wild Weasels (the planes from George AFB) to C-130 low-level electronic interference planes. The new planes would have little use for a new, high-tech bombing range, thus undermining a key argument by the Air Force that the range was being proposed primarily for the Air National Guard.

Later in April, U.S. Sen. Larry Craig called Bob DiGrazia, looking for compromise. DiGrazia said he would make no deals. Craig asked what opponents would think of simply expanding the Saylor Creek range, as suggested by the BLM. DiGrazia said it would probably be well received. Craig said he was under intense pressure by the Congressional Sportsmen's Caucus, National Rifle Association (of which he was and is a national board member), the Safari Club International, and others. Craig confirmed that C-130s were to replace the F-4s at Gowen Field, which they eventually did, along with ground-attack A-10 Warthogs—neither of which needed an electronic combat range.

To the range opponents in early 1994, it began to look like the Air Force, with the letter from Collier in hand, would start work on the range that summer. But the Air Force could not proceed without completing the final environmental impact statement.

The document, due out in April, remained on hold until Babbitt could take a position on the land exchange. The Interior Department was discussing the issue with the White House. And a White House official questioned Idaho Atty. Gen. Larry EchoHawk about the proposal. Governor Andrus was surprised to learn that the White House would make the call on the range that he had expected the secretaries of the interior and the Air Force to make. The White House wanted to make sure the Interior Department had looked at all the options. Administration officials were concerned about the effects of the range on the Shoshone-Paiute Indians, loss of wetlands, threats to wildlife, and damage to archaeological sites.

And then, through a mutual acquaintance, Bob Stevens contacted Theresa Heinz, wife of the late Sen. John Heinz. Stevens met with her, briefed her on the issue, and she became interested. She still had some connections at the White House. Over dinner one evening with Al and Tipper Gore, Heinz, who also helped pay for some of the efforts to fight the range, urged Gore to kill the proposal, Stevens said. Later in 1994, at a political fund-raiser that had nothing to do with the range, Rick Johnson of the Sierra Club met First Lady Hillary Clinton. When he was introduced, one of many lobbyists, he asked her offhand if she had heard of the Owyhee Canyonlands and the bombing range proposal.

She had.

When the First Lady knows about a bombing range in a remote corner of Idaho, "that's penetration," Johnson said. And that was the work of Theresa Heinz and, indirectly, of Bob Stevens.

The final impact statement and Air Force Secretary Sheila Widnall's decision on the range were expected by June 1994. At the same time, the Interior Department was to make a decision on the state's land exchange. BLM officials said Interior Secretary Bruce Babbitt had told them not to be too critical in their comments on the proposal, that President Clinton would make the final decision on the range.

But Babbitt's own agency had already raised serious questions about noise levels, aircraft accidents, the effects of fire, conclusions in the impact statement unsubstantiated by facts, the loss of wetlands, and inadequate evaluation of the cumulative effects in the impact statement. The document did not meet the requirements of the National Historic Preservation Act of 1966, and it did not meet the trust responsibilities of the United States for American Indians.[6] A historic 1866 battlefield that the Shoshone-Paiutes considered sacred would lie within the proposed range, and BLM officials recognized that it would not be possible to mitigate the characteristics that made the site significant.

President Clinton, in a February 1994 executive order, had outlined the United States government's trust responsibilities with American Indians. The BLM was obligated to protect areas with important traditional and religious values for them. Fulfilling that responsibility would make it impossible for the BLM to also support the Air Force's preferred alternative as outlined in the impact statement. If the land exchange were to go ahead, additional public comment and a full review by the Advisory Council on Historic Preservation would be required. The BLM recognized that its decision on the land exchange must be based on this review in addition to the impact statement. But a consensus was not likely. It was hard to see that the BLM could approve the proposal without serious changes. And it was unlikely that the Shoshone-Paiutes would change their position.

At issue was a rocky butte that rises more than one hundred feet above the spring-fed valley deep in the desert. The butte would have been at the heart of the north range—an area known to white people as Dickshooter Ridge, known to the residents of the Duck Valley Indian Reservation as "Sihwiyo," or Willows Growing All in a Row. Tribal Chair Lindsey Manning said his people had been coming to the site for more than a hundred years to conduct religious ceremonies, honoring their dead ancestors buried at the foot of the butte.

There in 1866, a group of Indians, camped along a small stream below the butte, fought off a unit of volunteer cavalry. Impatient with the U.S. Army, miners in Silver City to the north had set a bounty on scalps—$100 for each man, $50 for each woman, and $25 for "everything in the shape of an Indian under 12 years of age."[7] They equipped parties to go out and hunt humans. From the rocks atop the butte, the natives were able to keep the white "Indian hunters" from the water. The whites eventually withdrew. Dead from both sides are buried at the foot of the butte.

Though normally reluctant to reveal sacred sites to outsiders for fear of desecration, a group of Shoshone-Paiutes in September 1993 took some white visitors to the butte to explain its significance and importance.

The group of about two dozen people formed a circle around Butch Russell, an American Indian medicine man from Salt Lake City. The western horizon had just swallowed the sun. With a shrill and plaintive tone, he called to the

Battle Creek. *Photo by the author.*

spirits on a whistle made from the leg bone of an eagle. Kneeling before a ceremonial fire, he threw on a pinch of incense and fanned the fragrant tendrils in the four directions with the wing of a hawk. He intoned a prayer. Manning beat a simple rhythm on a hand-held drum as he and Terry Gibson sang a traditional song. The medicine man unwrapped his hand-carved, two-foot long ceremonial pipe. He tossed a pinch of tobacco in the fire for the spirits, a pinch on the ground for the ancestors, and packed a third into the pipe, repeating the process until the pipe was full. Manning lit the pipe and passed it to each person in the circle. No one spoke. The spirits of ancient warriors danced in the lavender twilight shadows among the rocks and sagebrush.

Early the following year, tribal members took a group of Air Force and Air National Guard officers to visit the sacred battleground. They participated in a similar ceremony honoring the fallen warriors. The officers were moved. An Air Force general told the Shoshone-Paiutes that he hoped they would have that place to pray in for a thousand years.

The tribes remained adamantly opposed to the proposal. Bombs would poison the water in the desert; the water would poison the animals that drank it; and there would be nothing out there, said Elaine Egan of the Duck Valley Indian Reservation. Her tribal name was Owyhee Kingie, and she was one of many Shoshone-Paiutes who objected to the proposed range just west of the 290,000-acre reservation. Egan, in her eighties and one of the tribes' elders, said the reservation was given to the tribes by the white man's government.

The reservation was established on April 16, 1877. The Western Shoshone were joined there in the mid-1880s by a group of Northern Paiutes. Between 1880 and the early 1900s, other groups of Shoshones and Paiutes moved to the reservation. Today about twelve hundred live there and make their living, mostly from ranching and farming. If the Air Force wanted to put in the bombing range and fly low over the reservation, it should be Congress, not Air Force or Idaho officials, that made the decision, Egan said.

Sherry Crutcher, a Head Start teacher on the reservation, said low-flying planes often scared many of the twenty children in her class. The jets flew so low that they said they could feel the heat from the engines, and the children were frightened by the noise.

Tribal member Kenneth Harney said the bombing range would make it hard to get the buckskin the tribes use to dress Indians when they die. The white man took over the North American continent with little regard for the people who already lived here. Now there is not much land left for the those people, and even that they cannot leave alone, he lamented.

The ranchers near the range planned to move away in exchange for millions of dollars. The tribes did not have that choice. They believed the area had been set aside perpetually for hunting and other traditional uses, and they wanted the range located somewhere else, tribal officials said. They asked that no supersonic flights be allowed over the reservation and that no flights be allowed lower than two thousand feet above the reservation. They also asked the Air Force not to use

Tribal attorney Dan Press and Lindsey Manning on horseback along Battle Creek. *Photo by the author.*

flares over the reservation or over the proposed range and not to use chaff where it would drift onto the reservation. There were sufficient existing areas through-out the United States to train the military without spending national resources to destroy their homeland, the tribes said. The proposal to situate the range adja-cent to the reservation was made without consulting the tribes, and they consid-ered that intolerable.

At a hearing on the reservation, a group of women cornered Governor Andrus' Air Force liaison Dave Jett, firing questions at him. Not satisfied with his answers, the women closed in and their questions intensified. Jett answered with polite thank you's. The women were angered when he appeared not to take their concerns about sacred sites seriously.

"How would you feel if we went to Mountain Home and bombed all your churches?" they asked him. "This is sacred ground to all us Native Americans. It's the same thing."

Manning said the tribes' concerns about the effects of repeated sonic booms and low-level overflights on the health of the Duck Valley residents were swept aside by Governor Andrus and the Air Force. Those on the reservation were not against the Air Force or Mountain Home, but they wanted the jets to stay away from the reservation and their sacred grounds, Manning said.

"Idaho gets the economic benefit. We get the shaft," he said. "We're not get-ting a red penny."[8]

But the reservation did eventually get some help. In February 1994, the Air Force agreed to provide increased medical treatment for tribal members at the base hospital in Mountain Home. Tribal members received emergency care, dental, and psychological services at the fifteen-bed reservation hospital. They had to go to Boise hospitals for surgery. The agreement included treatment at the twenty-one-bed base hospital paid for by the Indian Health Service, training for emergency medical technicians, doctors and nurses at the reservation, backup ambulance service, and technical advice on improving radio communications. The tribal members insisted the agreement would not dampen their opposition to the proposed range. And the Air Force insisted the agreement was not a way to buy them off.

Chapter 11 / Legal Challenges

T HE CONTROVERSY OVER the range continued to heat up during the summer of 1994. The environmental impact statement was not released as expected in June. National environmental groups were working on riders they hoped to attach to military appropriation bills to halt all military land exchanges without a clearly demonstrated need. The Foundation for North American Wild Sheep and The Wilderness Society highlighted the bombing range issue at a June 17 congressional hearing on a House bill sponsored by Minnesota Democrat Bruce Vento, chair of the House Natural Resources subcommittee on public lands. Vento's bill would have required congressional approval for military expansion on public lands for National Guard units.

June came and went. The continuing delay caused concern among supporters of the range and hope among opponents. Officials from the White House and the Defense and Interior departments met in mid-July to discuss the range; no news of that discussion leaked out. Ready or not, U.S. Sen. Dirk Kempthorne of Idaho wanted the document released. Air Force Secretary Sheila Widnall supported the proposal, and the Defense Department reportedly was ready to sign off on it. But the Interior Department, which had the final say over the land swap, was holding things up. The Shoshone-Paiutes' historical and religious use of the area could not be ignored.[1] Interior Secretary Bruce Babbitt had still not taken a side, and Defense Secretary William Perry had not endorsed the proposal.

Gov. Cecil Andrus repeatedly blasted the Clinton administration for dragging its heels. He said he would go directly to Babbitt, but Babbitt refused to be nailed down on when a decision would be made. The BLM felt pressured, but officials did not like the proposal or the land swap. Babbitt told Andrus that a decision could be delayed until the end of the year—after the November election. Some thought the delay was designed to benefit Idaho gubernatorial favorite state Atty. Gen. Larry EchoHawk, who the Democrats expected to replace Andrus when he retired from public office at the end of 1994. Andrus accused the Clinton administration of playing politics and allowing environmentalists to write military policy. But he did not directly blame fellow Democrat Bill Clinton, a former colleague from the National Governor's Conference. He blamed the president's bureaucratic underlings—environmental purists, who left no room for compromise, he said.

Like the thunderheads that rise over the high desert, spawned by summer heat, opposition continued to grow across the country—from New York to San Francisco, Boise to Washington, D.C. Increasingly, the delay in the impact statement looked like a political and legal tug-of-war. But the proposal would suffer two serious blows in early fall. Widnall and a federal judge would both conclude that the impact statement was inadequate.

Meanwhile, two stalwarts of the opposition, Bob Stevens and Herb Meyr, convinced the Center for Defense Information, a Pentagon watchdog group, to prepare a documentary on the range issue. Stevens took a reporter and cameraman up in his plane for a look at the Owyhee Canyonlands. The resulting documentary fanned the flames of controversy, piquing the interest of lawmakers and renewing doubt about the project among environmentalists and Hollywood actor-activists. Congressman Vento said the proposal was a deliberate attempt to sidestep the Engle Act. Actor Paul Newman questioned the idea of risking the natural area to build a range for which the Air Force had failed to show a need. Sens. Harry Reid of Nevada, Daniel Inouye of Hawaii, and Ben Nighthorse of Colorado said the range was ill-advised and violated federal law. The proposal was not well thought out, and the Air Force appeared to be trying to reach a hasty decision on the idea developed by the state of Idaho.

By late August, Andrus had received a preliminary review copy of the final environmental impact statement. But it was unsigned and not yet complete. It did not answer the concerns of the American Indians or of Fish and Game, hunters, hikers, river runners, or environmentalists. It said nothing about avoiding sensitive wildlife areas or mitigation. Increasingly frustrated, Andrus again threatened to go directly to Clinton, and if White House staff kept him away, he would go to the national press. But so far his contacts in the White House had failed to get him in. He was unable even to get Vice President Al Gore to discuss the issue with him.

On September 9, 1994, Bob DiGrazia of the Foundation for North American Wild Sheep urged Kathleen McGinty, head of the White House Office of Environmental Affairs, to tell the president that Idaho was trying to dictate public lands and national defense policy with the land exchange. That same month, about thirty members of Congress asked Clinton not to act on the range proposal without conferring with Congress. The members said they were responding to concerns of fourteen environmental groups and the Shoshone-Paiutes.

Later in September, forty members of Congress noted their reservations about the range to Congressman John P. Murtha, chair of the House Appropriations Subcommittee on Defense. The concerns were the same. They urged Murtha to adopt the Senate appropriations report that blocked any expenditures until the secretary of defense explained the need for the range and why existing facilities were inadequate. The report also asked the Defense and Interior departments to explain why the Engle Act should not apply—and how aboriginal rights would be recognized.

With advice from the White House, DiGrazia said, the Foundation for North American Wild Sheep, on September 29, plopped down a twenty-five thousand dollar donation from the Aldo Leopold Society in San Francisco for a full-page ad on the back of the front section of the *New York Times.*

Back in Idaho, Andrus was pushing the Idaho Land Board—of which he was chair—to approve the land swap at its October 11 meeting. Most board members supported the swap, with some conditions. EchoHawk said he wanted to see all the facts, including the final impact statement, before voting on the exchange. Some skeptics suggested it was a delaying tactic to avoid taking a stance until after the election. The other three members, Schools Superintendent Jerry Evans, Secretary of State Pete Cenarrusa, and Auditor J.D. Williams, supported the trade. U.S. Congressman Mike Crapo of Idaho, after visiting the Duck Valley Indian Reservation and hearing the concerns of its residents, urged the Land Board to approve the land exchange. But it never came to a vote.

On October 4, 1994, Secretary Widnall announced that the final impact statement would not be released. Instead, the Air Force would continue to study alternatives in an effort to reduce environmental effects and to relieve the concerns of the Shoshone-Paiutes. Air Force officials had developed a comprehensive mitigation plan, she told Andrus. But because of the significant changes to the proposal in that plan, and to consider other alternatives, the Air Force would release a supplemental environmental impact statement, and with it, a new round of public hearings.[2]

Andrus was not fooled. He saw clearly that the proposal was dead. "The blue-suiters marched us down the primrose path, and now, at the eleventh hour, they have walked off and left us," he said. He later stated that the announcement was his second great disappointment, behind the Fish and Game Commission pulling its support.[3]

Opponents, reluctant to celebrate victory prematurely, welcomed the announcement. But nobody was declaring the project dead. "This thing is gut-shot and both kneecaps have been hit, but it's still wiggling," Bob Stevens said. "We're not giving up until it stops wiggling."[4]

The Land Board put the land swap on the shelf. Department of Lands Director Stan Hamilton didn't expect the Air Force to come back with another proposal for several years. Andrus gave them a year and a half—at least. But the Air Force was not giving up. One officer spoke of plans to pursue a modified proposal that would place target areas for dummy bombs on the southern half of the range. The northern half would have only moveable emitters that would be attacked electronically; no bombs would be dropped.

Andrus, GOP Sens. Craig and Kempthorne, Congressman Crapo, and other supporters lashed out in bipartisan outrage at what they termed political maneuvering by insiders at the White House who wanted to kill the proposal. But EchoHawk noted that the Air Force was forced to take another look at the range simply because the impact statement was inadequate. It would have to be redone, not because of politics but for valid reasons that had come up long ago—reasons carefully documented by range opponents and the experts they had hired, the BLM, the Idaho Department of Fish and Game, and the Shoshone-Paiutes. The range proposal did not have the broad support necessary to move it along on a fast track.[5]

Three days after Widnall's announcement, a federal judge reached the same conclusion. On October 7, 1994, Federal Magistrate Mikel H. Williams ruled that the evidence clearly showed the composite wing and the proposed range were "inextricably intertwined."[6] He recommended that the Air Force start over and complete a single environmental impact statement, analyzing the effects of the state range proposal and the composite wing together. "Not to require this would permit dividing a project into multiple actions, each of which individually has an insignificant environmental impact, but which collectively have a substantial impact."[7] Failure to include the range and the composite wing in the same document was an abuse of discretion and a violation of federal environmental law, he ruled.

The ruling stemmed from the lawsuit Murray D. Feldman had filed on behalf of the Greater Owyhee Legal Defense back in 1992. From Feldman's fifteenth floor office in downtown Boise, the Owyhee Mountains hid the Owyhee Canyonlands to the south and east. But he could clearly see that the Air Force's 1992 decision to pursue the state's range proposal on that high desert plateau violated both the letter and the spirit of federal environmental law. Air Force planners should have considered the composite wing and the range proposal in the same impact statement. They should have looked at other range locations and other ways of meeting training needs. They should have considered a smaller composite wing and alternatives to the use of chaff and flares and to the airspace changes and supersonic operations. Without considering such alternatives, they had no basis for their conclusion that the range was operationally and environmentally suitable. In addition, Air Force planners did not sufficiently establish the need for the range as required by law. And though they analyzed the economic benefits of the changes to Mountain Home and Elmore County, they failed to analyze the environmental costs to Owyhee County.[8]

The court at the same time was considering lawsuits filed by the Shoshone-Paiutes and Richard and Marie Owen, who owned the Big Spring Ranch for which the state's original proposal—the Big Springs Training Range—had been named. Though it would have bordered the range, none of their private land would have been purchased for the range, and compensation plans did not include the Owens. The state adopted the name of their ranch without consulting them, Richard Owen said. They said the Air Force must prepare a complete environmental impact statement on the proposed range before officials could find it environmentally suitable. The Shoshone-Paiutes sued to protect religious and historic sites. Federal law gives American Indians the right to believe and practice traditional religions. And the Air Force admitted that its proposed operations would affect native ceremonies and sacred sites. Without reconciling the conflict between the proposal and the American Indian Religious Freedom Act, the document violated the law, Feldman said.

In his most telling point, Feldman argued that federal law required that the cumulative effects of related or connected proposals be considered in the same impact statement. But the Air Force failed to consider the cumulative effects of the range and the composite wing together. The decision to pursue the state range proposal was directly related to the composite wing because it created the need for a tactical and electronic combat range. Air Force officials knew at the time they wanted additional training space, but they had told the Base Closure Commission that existing ranges would accommodate the composite wing. They had tied themselves a Gordian knot. To now say in court that the wing needed the proposed range would have amounted to an admission that they lied to the commission and to Congress. And it would have supported GOLD's contention that both should have been analyzed in the same impact statement. They could not write an impact statement that asserted the range was not needed either; that would have made the effort pointless. Air Force officials instead hung the proposal on their contention that the range would improve training in Idaho—a point they failed to prove.[9]

GOLD's lawsuit asked that the Air Force halt its proposals to establish the composite wing, expand military airspace, add supersonic operations, change the mission of the Idaho Air National Guard at Gowen Field in Boise, and study the state's range proposal. All should have been considered in a single impact statement. Feldman charged that the Air Force had separated the environmental reviews to reduce the apparent effects of proposed changes to expedite the process. Decision-makers and the public were not getting the whole picture. Taken individually, most of the changes in the Air Guard operations and in the airspace would have had little effect; the new composite wing by itself would have had some additional effect; but the proposed range was where those effects would show up together as potentially significant. And that part was left out of the 1992 impact statement. The Air Force had created the wing and the airspace for its three-million-acre electronic, tactical, and conventional battle complex, examined them in one document, and deferred analyzing the environmental effects until the second. Feldman argued that considering these issues separately constituted an "impermissible segmentation of the overall Air Force proposal that masks the true environmental impact of the proposed actions and prevents the public and agency decision-makers from obtaining full disclosure of the environmental impacts of the proposed actions."

Assistant U.S. Atty. Marc Haws, representing the Air Force, argued that the wing and the range were not connected, that the wing did not need the range, and that the range was a state proposal and too vague at the time to be included in the same impact statement with the wing. And because the bombing range was a state action, federal law requiring an environmental impact statement did not apply.

In January 1993, Haws had asked the judge to dismiss the tribes' lawsuit, and Feldman figured GOLD was next. In April, Haws did just that. The Air Force wanted to resolve the lawsuit in order to release the second impact statement on the range proposal. But to Feldman it was a tactical blunder. Haws had asked the judge to dismiss the lawsuit without submitting a complete administrative record—and that was "like asking the judge to look at the case with his eyes closed." Feldman noted that the record did not include a number of studies, reports, land management plans, and other documents cited in the impact statement. He maintained that the Air Force wrongfully tried to withhold these documents from the court and from the public.

Haws insisted the documents, though listed in the impact statement, were not actually used by the decision-makers and therefore were not legally part of the record. The fact that they were cited made them part of the record, Feldman countered. Either the Air Force violated the law by failing to properly consider all the available information relevant to the proposal, or the documents were in fact considered and therefore a part of the record that must be filed with the court. The Air Force argued that the court's review of the case should be based solely on the impact statement and documents already released. But in earlier arguments in 1992, Air Force lawyers had said that a complete administrative record would benefit their case because the case probably would be decided based entirely on a review of that record.[10]

Judge Williams sided with Feldman and ruled in January 1994 that the Air Force could not withhold any documents that were part of the administrative record. The case would indeed be decided on the complete record, and that record clearly showed the connection between the wing and the range proposal. Deputy Assistant Secretary of the Air Force Gary Vest had said the proposal offered an opportunity to develop a "training infrastructure to accommodate the evolving and future needs of the Air Force's composite wing."[11]

Air Force officials also had said that full consideration of the alternatives for the range was not appropriate in the 1992 impact statement because they had not yet decided to pursue the range. Feldman argued that the whole idea of the impact statement was to analyze the environmental effects of a proposal before a decision was made to pursue that proposal. The Air Force's later impact statement on the range proposal carefully explained why the Saylor Creek range was inadequate, while Haws argued that the composite wing did not need the proposed range to operate.

Haws further argued that the move of F-4s from George Air Force Base in California to the Air Guard at Boise's Gowen Field formed the basis for the Air Force considering the bombing range proposed by Governor Andrus. And that was what was being studied in the impact statement, Haws said. But the Air Guard in a 1991 environmental assessment had declared that the F-4s did not need any additional training facilities. The Saylor Creek Range would satisfy the requirements of the F-4 aircraft, and any new range or airspace would only en-hance training, said Ronald Watson of the Air National Guard.[12]

Andrus had proposed the range as a way of making Idaho more attractive to the Air Force. And his offer may have helped the Air Force decide to establish the composite wing at Mountain Home. Air Force brass, after reviewing the pro-posal, decided it should be actively pursued because it would "primarily support active Air Force forces."[13] In August 1990, Air Force Secretary Donald Rice had designated Gary Vest as the single point of contact for the Defense Department and the Air Force to consider Andrus' plans to accommodate Air Force training needs in Idaho. Vest's assignment was to work on a solution that would satisfy the training needs of the Air Force and the concerns of ranchers, environmental-ists, and other opponents of the failed Saylor Creek expansion. Vest in turn ap-pointed an officer to develop generic needs for a practice bombing range. But particulars, including land area and airspace requirements, depended on aircraft and weapons systems, tactics, and the degree of readiness—all influenced by the composite wing structure, underscoring the interdependence of the wing and the range. One of the fundamental reasons for establishing the wing was to pro-vide specialized training to a fighting force composed of several types of aircraft. To do that, the wing needed adequate airspace, supersonic operations, a conven-tional bombing range, an electronic combat range, the ability to perform defen-sive countermeasures, and electronic radar threat emitters. The state range would include all of those elements.

Air Force officials had authorized a study of the area in September 1991 and hired a contractor to develop detailed range layouts with precise target area boundaries. The Air Force knew it wanted an expanded range. Establishing the composite wing had not done away with the inadequacies of the Saylor Creek range. By combining improvements at Saylor Creek with the range proposal, Mountain Home would have excellent composite force training ranges, the Air Force had said in the spring of 1991. To claim that the range was not a federal proposal, and that it should not be considered in the same impact statement as the wing, was "simply preposterous," Feldman said. The link between the two was clear. The range was designed to accommodate the needs created by the composite wing.[14] The two were in fact intimately connected until the Air Force intentionally separated them to avoid letting environmental concerns over the range delay the "beddown" of the composite wing.

Judge Williams agreed with Feldman. And in October 1994, for the second time in a week, the range proposal suffered a serious blow. Williams' recommended ruling would be sent to U.S. District Judge Harold L. Ryan, who would make the fi-nal ruling after reviewing the record. The case was eventually taken over by District Judge Edward J. Lodge. With Williams' blow following Widnall's announce-ment, most were certain the proposal was dead.

Victories are hard to come by in his business, Craig Gehrke said. But when he got the call from Kathleen McGinty in October that the Air Force was withdrawing the proposal, he knew the coalition had won that round. "Oh God, that was sweet."[15]

Chapter 12 / Critical Mass

RANGE OPPONENTS WERE GLEEFUL when they heard President Clinton had ripped up the Air Force impact statement. Reacting to public concerns and comments from environmental groups and tribal leaders, Clinton had sent the Defense Department back to rethink the proposal in November 1994.[1] But Governor Andrus contended that the impact statement never got to the president but rather was stopped by White House environmental policy head "Katie McGinty and her wrecking crew." Andrus felt that while Clinton had good environmental credentials, some of the younger members of the White House staff did not; most of them had never been west of the Mississippi River. The problem was that they were trying to go too far; they were being environmental idealists. They wanted it all, and they blocked anything less—you cannot compromise that way, Andrus said.[2]

By the time word of the long-awaited impact statement reached Idaho, however, onion farmer and former state legislator Phil Batt had won the election over Andrus' expected successor, Atty. Gen. Larry EchoHawk. When Batt took over the governor's office in January 1995, the range was still in limbo, but he said he would pursue it and wanted it approved by the middle of his first year in office. But Batt was not Andrus, nor did he have Andrus' skill as a politician or his clout in Washington. Batt urged the state legislature to endorse the range proposal, a showy but largely ineffectual move.

The supplemental impact statement that Air Force Secretary Sheila Widnall had said would clear up all outstanding concerns had been due out in December 1994, but by January it still had not been released. Air Force officials told Batt that they had developed significant changes and were considering others that would improve training while responding to public concerns. The new document would ensure these changes were properly assessed. Air Force officials said they had learned they had to resolve local community and American Indians' concerns "with the same thoroughness we use to address combat readiness."[3] They asked Batt to work with them to complete the supplemental impact statement and reach a final decision.

Air Force officials seemed to be stringing Batt along with vague promises while Idaho's congressional delegation was trying to win support for the moribund proposal in Congress. Despite a federal district judge's approval of a recommended ruling against the environmental impact statement, Air Force officials continued to reassure Idaho. Then in a surprise reversal, the Air Force said it had already given up on the state's proposal, and a Defense Department audit deemed the proposed range redundant. Still, the proposal refused to die easily.

In February 1995, Defense Secretary William Perry told the Senate Armed Services Committee that he expected the range issue to be resolved soon. The rumor, spread earlier by an Air Force official, resurfaced that the modified range would place target areas for dummy bombs on only the southern half of the range. The northern half would have only moveable electronic emitters. Rumors also surfaced about a deal with the Shoshone-Paiute tribes that would limit over-flights of the reservation.

In March, Batt appeared worried that the proposal was getting away from him. He reminded Perry and Widnall that it was, after all, a state proposal, and the state ought to retain control of the alternatives being developed. State officials still had not heard any details of the proposed changes. He requested a meeting with Perry and Widnall and the Idaho congressional delegation—no later than April 7—in his Boise office. He was troubled by the delay in public scrutiny of the alternatives and mitigation measures, he said, and he wanted to see the alternatives "shaped into a range that would accommodate the needs of military training with proper environmental and cultural protection."[4] Widnall assured Batt that the Air Force still was committed to the range proposal and still committed to working with the state.

Dave Jett, who still was working as the governor's liaison, noted that state officials were working on range funding and management, a range management plan, a cultural resource management plan, recreation, a road agreement, and the land exchange so that whenever they got the details of the supplemental impact statement, they would be ready to complete work on the range.[5] But first the modified range plan had to be approved by Vice President Gore—that meant Theresa Heinz, Jett noted. The BLM would have to approve field studies of range locations proposed by range opponents. And the public comment period for the supplemental impact statement would give range opponents one more opportunity "to wave the flag to save the world from the big bad (Department of Defense)."[6] The plan also required agreement of the Interior Department and the Shoshone-Paiutes, but any conflict between the Interior and Defense departments would have to go to the White House for settlement. And that meant Kathleen McGinty would make the decision—against the range. During the last round of hearings, eighty percent of comments had supported the range, but the White House chose to listen to the thirteen percent who opposed it, Jett claimed.

The longer and the more money it took, the less attractive the range proposal would be to the Air Force, Jett said. Dollar and political costs soon would outweigh the value of the range. And if the Air Force started looking at other areas, it might lose the support of Governor Batt and Idaho's congressional delegation. The governor needed to demand that the Air Force consult with the BLM and the state to get the details of the supplemental impact statement before it was presented to the public.[7]

Meanwhile, Congressman Mike Crapo and Congresswoman Helen Chenoweth of Idaho were trying to bolster support for the proposal in Congress. They had invited U.S. Congressmen Duncan Hunter and Randy "Duke"

Cunningham—both California Republicans—to tour the Owyhee desert in a pair of U.S. Army Blackhawk helicopters in late April 1995.

Following the tour, Hunter and Cunningham vowed to force the completion of the proposed range, insisting pilot lives were at stake. Hunter, third-ranking member of the House National Security Committee, said he would push for a hearing before his committee and ask it to subpoena the impact statement that had been stalled since the previous October. Cunningham wanted to use his reputation as a top naval flier in the Vietnam War—one of only two American flying aces in that war—in support of the proposed range.

But their claim that the range was needed for national defense was untrue, Herb Meyr said. Mountain Home had made it through the most recent rounds of base closures, and the wing had won top honors in Air Force competitions—without the state's range. Air Force officers had pronounced the wing the most combat ready in the world, and the base one of the top bases in the Air Force. Mountain Home pilots would be no better trained at an Idaho range than they would be at Utah or Nevada ranges. Besides, national defense issues should be considered in Congress—not in the Idaho governor's office, Meyr and others noted. Air Force leaders would say whatever they could, to whomever they could, to get this thing going, Craig Gehrke said.

But the two congressmen had not come to Idaho simply to drum up support for the failing range proposal. Earlier in April, according to Fish and Game records, Chenoweth had paid for Idaho non-resident hunting licenses and non-resident turkey tags for Cunningham and Hunter—a total of $274—with money from her own campaign funds.

The battle appeared over on May 9, 1995, when U.S. District Court Judge Edward J. Lodge, who had taken over the case from Judge Harold L. Ryan, approved Magistrate Mikel Williams' recommendations of the previous October. Lodge ordered the Air Force to conduct a combined environmental impact statement that covered both the composite wing and the range in a single document. In his ruling, Lodge swept aside Air Force objections. Air Force lawyers had argued that the composite wing at Mountain Home Air Force Base had successfully trained without the range and therefore the two were not connected. They also argued that the Air Force was bound by law to establish the wing at Mountain Home—range or no range—by the Base Realignment and Closure Commission's decision, which became law when the president signed it in July 1991. The Air Force never once considered the state proposal as a requirement for establishing the wing. The Air Force did not need the range, but the training that the range would provide was a requirement.

But those were the same arguments that Lodge already had ruled against. Air Force lawyers did not offer anything new upon which to base a reversal of Williams' report, Lodge said. The Air Force's version of the facts contradicted the administrative record and facts to which the Air Force already had agreed. "The objections primarily attempt to refashion the facts in an unsuccessful effort to avoid the findings and conclusions of the magistrate," Lodge wrote.[8]

Williams had also recommended that both parties in the lawsuit discuss a proposed injunction. GOLD had sought to keep the Air Force from completing the proposals covered in the composite wing impact statement. If the two sides could not agree, then the issue would be returned to the court to be decided. If the composite wing already was training successfully elsewhere, the Air Force would suffer little harm and an injunction would essentially maintain the status quo until a combined environmental impact statement could be completed. Lodge agreed. Everyone thought the range proposal was finally dead.

But it was still wiggling.

On the same day Lodge issued his ruling, Air Force Secretary Widnall assured Governor Batt of her continuing efforts to complete the supplemental impact statement on the proposed range. The Air Force was working with the Shoshone-Paiutes and other opponents and making significant progress, she reported on May 9. "The Air Force is committed to pursuing this proposal,"[9] she wrote. Her letter included a copy of a March 24 letter from the Shoshone-Paiutes, telling Air Force Undersecretary Rudy DeLeon that the range proposal was unacceptable. The letter included a map and a set of conditions under which the tribes would be willing to work to develop a range. But the areas on the map not acceptable to the Shoshone-Paiutes for any range—drop or no-drop— would rule out the state's proposal entirely. It should have raised suspicions. The tribes wanted no sonic booms, no flares or chaff or flights below twenty thousand feet above the reservation or over sacred sites, no flights over the town of Owyhee, no mobile emitters near sacred sites.

On the following day, May 10, the Idaho delegation and Governor Batt, in a meeting with Widnall and military and BLM officials, received word from Air Force, Interior Department, and White House officials that the Clinton administration had changed course and was now behind the range proposal. Thomas Jensen, associate director for natural resources in the White House's Council on Environmental Quality, announced that there was "100 percent complete support from the White House" for developing the range.[10] The announcement was met with praise from the Idaho delegation and Batt.

"The White House is now realizing their anti-West style is not getting them anywhere, and it is a real shift on their part in realizing the value of the training range," Senator Craig said. Senator Kempthorne added: "We've never felt there has not been full commitment from the Air Force. What was lacking was a commitment from the White House." Kempthorne served on the Senate Armed Services Committee and Craig sat on the Energy and Natural Resources Committee—both key committees to the range proposal.

The time had come to proceed on the state's proposed range, Widnall said following the meeting. She and the delegation would review Judge Lodge's ruling, and they would work with the state, the Air Force, and the court to reach a satisfactory resolution. The ruling was seen as a delay in the process, but not as an insurmountable problem. Congressman Crapo characterized the ruling as procedural. Craig considered it significant, but without any sense of alarm; the judge's concerns could be worked out in the supplemental impact statement, he said. Like the others, he came away from the meeting feeling that everyone was trying to resolve the remaining range issues as quickly as possible. Final resolution was at hand, he said.[11]

But in a sworn statement two weeks later, on May 23, Air Force Brig. Gen. Michael J. McCarthy said the Air Force had decided to drop the Idaho proposal before Lodge's May 9 ruling. McCarthy at the time was in charge of ranges and airspace and was responsible for consideration of the Idaho range proposal. He said that after studies and consultations with the Interior Department, the state of Idaho, the Shoshone-Paiute tribes, and others, the Air Force had decided it would no longer pursue the state's range proposal. Two weeks after the fact, McCarthy claimed the Air Force had reached this conclusion before the judge had issued his ruling. As a result, there would be no range proposal for the Air Force to consider in a combined impact statement, he said. Despite Secretary Widnall's reassurances, Air Force officials now said they had stopped work on the supplemental impact statement. Instead they would spend the next few months talking over other possibilities with Interior Department and Idaho officials. McCarthy also insisted that the Idaho delegation and Governor Batt had been told at the May 10 meeting that the Air Force had decided to drop the state's range proposal. The only evidence to support McCarthy's assertions, however, was his own statement two weeks after the meeting.

Now what? Was this a ploy to get out of redoing the impact statement? Or was it a way of untangling the next range proposal that most were sure would follow—a range proposal apparently already in the works, despite contrary statements from some Air Force officials?

Based on McCarthy's sworn testimony, the Air Force went back to court to seek relief from the order to conduct a combined environmental impact statement. It would be unnecessary and a waste of time to conduct an impact statement on a range the Air Force no longer was contemplating. Attorney Marc Haws asked that the Air Force be relieved from any injunction that might arise pending the completion of a combined impact statement. And he asked that the court vacate the entire order. Feldman argued that the judge still should require a combined impact statement for the composite wing and any other range proposal. The judge wanted to wait until the Air Force made another proposal.

Batt was a little surprised and disappointed with another apparent delay in the establishment of the state range.

Air Force officials told Bob DiGrazia of the Sheep Foundation that it no longer was working on any plans for expanded training facilities in Idaho. But should the Air Force hit upon an agreeable proposal for expanded training in Idaho, it would let everyone know, and it would conduct appropriate environmental reviews at that time, DiGrazia said.

Not long after the Air Force's surprise reversal, the Defense Department issued its own devastating blow to the state's range proposal. In an Audit Report, issued June 30, 1995, the Office of the Inspector General said existing ranges were adequate to train the composite wing at Mountain Home. Air Force officials had said the wing had to have the Idaho range to "provide quality, realistic training on a daily basis, close to home."[12] But the Air Force already had that, the audit noted. The Assistant Secretary of Defense for Strategy and Requirements argued that the existing Saylor Creek Range supported day-to-day training, and the associated airspace in Idaho accommodated some composite wing training during which pilots and aircrews of differing aircraft learned to work as a team. Additional training requiring more ground targets could be done at large ranges in Utah and Nevada. Training at the distant ranges would incorporate air tanker support, adding the realism to the mission that the Air Force said it wanted.

Auditors found that records for 1993 and 1994 showed the wing had flown to the Nevada and Utah ranges for about 8 percent of all training flights—not day-to-day, routine training. Auditors also found that the Air Force overstated flying time to the distant ranges by as much as thirty minutes. The Air Force had set a 150-nautical mile limit as the distance pilots could fly for routine, day-to-day training without aerial refueling. But the wing's records showed that about 80 percent of its flights to the Utah Test and Training Range—175 nautical miles away—were not refueled. The Air Force also complained about scheduling conflicts at the Utah range and at other ranges. But no records could be found that showed any training missions had been rescheduled or canceled because of schedule conflicts.

Despite the Air Force's lengthy rebuttal, it did not establish training requirements for the composite wing, nor did it show why the existing range could not meet the wing's training requirements. The auditors concluded that establishing the Idaho range would be an exception to the Defense Department's attempt to reduce its infrastructure. Such an exception should be approved only after close consideration of the need and other ways of meeting that need. Until a composite wing training plan and a corresponding analysis could be prepared and approved, no further funds should be expended on the range. A new range in Idaho would be redundant, the report concluded.

The Air Force, as expected, disagreed. The Idaho range would provide unique training scenarios that would enhance training. The Saylor Creek range could provide the basics, but it could not integrate simultaneous air-to-ground attackers in complex exercises, and some weapons releases were limited because of safety and airspace restrictions. The Air Force estimated it would cost $31.5 million to build and operate the range for six years to provide the 366th Wing with a few hours of training a month already adequately provided at other ranges.

By the middle of 1995, most thought the Owyhee Canyonlands were saved. Stevens packed up all his documents and gave them away. He and Meyr went fishing in British Columbia. Environmentalists went on to other issues—mostly. Ranchers went back to raising cattle. Andrus already had retired and returned to consulting on natural resource issues, founding the Andrus Center for Public Policy. He was glad he was no longer involved. In the end, the battle had boiled down to an athletic contest, Andrus said without apparent bitterness, sitting in his modest office at Boise State University. One day he told a friend that everyone owned a mistake. Dworshak Dam on the Clearwater River, approved despite its effects on elk habitat and opposition from the public and Fish and Game officials, was Sen. Frank Church's; the bombing range was Andrus'. But at the time, he thought he was doing the right thing.

He felt sold out by Fish and Game commissioners. And he felt let down by the Air Force "blue-suiters" who got the state involved in the first place. Air Force Secretary Widnall acquiesced to Kathleen McGinty, who played to the national environmental groups with little concern for folks in the West, Andrus said. "Katie McGinty put the screws to Widnall, and the Air Force backed off and left Cecil swinging in the breeze."[13] What do we have now? Jets flying low over the canyons with no restrictions. Andrus said his proposal would have kept them away from the canyons. And there was no real evidence the jets had caused the drop in sheep numbers; habitat conditions change, and it was in the middle of a number of dry years, he said. Besides, the sheep were more scared of the helicopters biologists use to count them. Tired of the controversy, Andrus no longer had any sympathy for the Air Force. But the state effort, though unsuccessful, showed the Air Force that the state was committed, and that helped keep the base open. "That was my motive." In that sense, Andrus' proposal was successful.

Part Three:
Enhanced Training

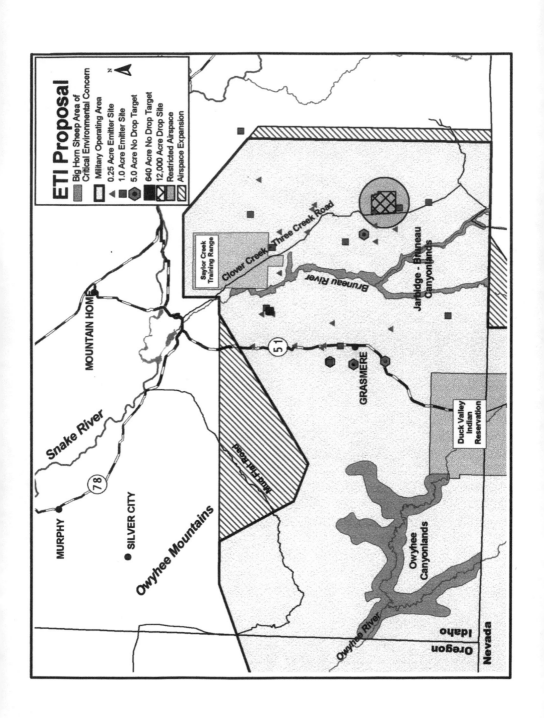

Chapter 13 / One More Time

T HE EFFORT TO EXPAND TRAINING in southern Idaho never slowed. Air Force officials did not want to lose what they had gained in the 1992 decision to establish the composite wing—increased airspace, supersonic operations, and the use of chaff and flares. The Air Force had washed its legal hands of the state's proposed range with a post-dated assertion that the proposal had been dropped before the judge's ruling that the range and the composite wing should be considered in the same impact statement. The ultimate result was that the Air Force would not have to abide by that decision.

Following the demise of the state's proposed range, Air Force officials in mid-1995 started meeting with representatives of the state, the Interior Department, the Duck Valley Indian Reservation, and others to "identify training opportunities in Idaho." But not everyone was invited. Congressman Mike Crapo—who at other times has said government decisions must be collaborative and include all those affected—said at the time that the state and the Interior Department would represent environmentalists. But that was unacceptable to the Owyhee Canyonlands Coalition. Air Force officials were repeating past mistakes of not including all interested parties in the planning, which had led to the drawn-out controversy of the past several years, said Lasha Johnston of The Wilderness Society's Idaho office.

"The proponents cannot sit behind closed doors and develop a proposal which is then delivered with their customary 'take it or leave it' attitude. This approach has not worked in the past and will not work this time." Johnston wrote to Air Force Secretary Sheila Widnall.[1] She urged Air Force and Defense Department officials to take an honest look at training needs and carefully study the costs and benefits of any proposed training facilities.

But in 1995 neither Johnston nor other range opponents knew that a deal already had been made in late 1994—a deal that all but assured the success of a subsequent range proposal. Not until 1998 did Brian Goller learn from a White House staffer of an agreement that ensured the support of Interior Secretary Bruce Babbitt and Council on Environmental Quality Director Kathleen McGinty, who both had opposed the state's proposal. The deal apparently was hammered out in October 1994, about the time Widnall shelved the impact statement on the state's range proposal and U.S. Magistrate Judge Mikel H. Williams ruled against it.[2] A deal would explain the sudden switch by the White House and the assurances without accompanying details from Widnall. Since then, Air Force officials had been looking for another place to site a training complex that would combine the existing Saylor Creek range with a new practice bombing range, electronically simulated bomb target areas, and an electronic combat range with mobile radar units simulating enemy air defenses in an area acceptable to the Shoshone-Paiute tribes.

In the privacy of congressional offices, Sens. Larry Craig and Dirk Kempthorne, Reps. Mike Crapo and Helen Chenoweth, and BLM, Interior Department, White House, and Air Force officials had agreed to follow a process that eventually resulted in a successful proposal for an Air Force range in Idaho. All agreed not to oppose the proposal if the process was followed and the concerns of all groups were resolved. But nobody asked the people of Idaho what they thought of this.

"Neither the media, the public, the tribes, nor range opponents knew of the backroom deal predetermining crucial support for an incipient range proposal," Goller said.[3]

Later in 1995, the Air Force announced its plans for a new range to be located in the same part of Owyhee County as the failed Saylor Creek expansion. The concept was the same, changed mainly in scope and size and without low-level supersonic flight or live ordnance. In addition, the new proposal would expand airspace over the Little Jacks Creek area, an area Fish and Game already had recommended against but which was important to the Air Force because it would extend the Saylor Creek operating area straight west and give pilots another approach to the range from that direction. Air Force officials said the expansion would reduce concentrations of fighters flying around the southern tip of the area.

The new proposal also would include one 12,000-acre practice bombing range, with a 300-acre impact zone; one 640-acre "no-drop" simulated target area; four 5-acre no-drop simulated target areas; and thirty electronic emitter sites to form an electronic combat range of more than 1 million acres. Up to forty planes at a time would train together from five hundred to fifty thousand feet above the ground. The proposed range would be spread from state Highway 51 to the Twin Falls County line, and from Three Creek to the Saylor Creek range—existing military airspace in northern Nevada would be expanded. Operations would include supersonic flight over ten thousand feet above the ground, with an average of two sonic booms per day, and the use of chaff and flares. On the proposed bombing range, pilots would drop only twenty-four-pound cast iron practice bombs, which would release non-flammable markers upon impact.

Air Force officials still said the range was not a requirement but that the wing needed it for efficient training, and they claimed that it was too expensive to fly to the ranges in Nevada and Utah every day—which they still would not have to do.

Air Force and Idaho officials set out to gain the cooperation of one group at a time. Once officials had agreement in concept with ranchers, they set out to talk with the Shoshone-Paiutes and other special interest groups.[4] Sen. Dirk

Kempthorne, who later praised the entire process as open and public, met with officials in closed meetings.

Chief of ranges and airspace at the Pentagon, Col. Fred Pease, told the *New York Times* that pilots would gain approximately twice as much training at a range in southern Idaho than if they had to commute to ranges in Nevada or Utah.[5] They could be training instead of flying during the time it took to fly to those ranges, he said. But he later admitted that was true only for the few times a month the pilots flew to those distant ranges. Idaho facilities would still provide about seventy-six percent of the wing's training. The proposed range would mean an additional fourteen percent could be done in Idaho. Pilots fly fifteen to twenty hours of training a month—fourteen percent of that equals two to three hours. He also admitted that a new Idaho range, while improving efficiency, would not eliminate the need to fly to the Nevada and Utah ranges. The remaining ten percent—training with live ordnance—still would have to be done at those out-of-state ranges. In addition, pilots still had to practice aerial refueling. And even without a new range, the wing had managed to train effectively.

So another round of hearings on a proposed range in Idaho began—this time, however, unencumbered by any realignment. But before the hearings got started, the Shoshone-Paiute tribes settled their earlier lawsuit separately in an August 1996 agreement with the Air Force. The Air Force had agreed, barring unspecified "compelling national security circumstances" that it would *not:*
- Fly supersonic missions over the Duck Valley Indian Reservation.
- Use flares at night or below twenty thousand feet above the ground during the day.
- Use chaff over the reservation.
- Fly at any altitude within five miles of the town of Owyhee.
- Fly below ten thousand feet above the ground over the reservation.

In addition the Air Force agreed to:
- Ask other airspace users to abide by the restrictions, though the Air Force would not be responsible for any non-compliance.
- Abide by a voluntary altitude restriction of fifteen thousand feet, which would be reviewed periodically.
- Pay the tribes' attorney fees of sixty-two thousand dollars.[6]

In December 1996, the U.S. District Court set aside a portion of its earlier ruling by not requiring the Air Force to complete a combined impact statement. The court denied attorney Murray Feldman's request for an injunction against composite wing operations until a new, legally sufficient impact statement was completed. The court said GOLD would have to wait until the Air Force filed a final decision on a proposed range before a lawsuit would be ripe.

But it was becoming obvious that the Air Force was not going to close the Mountain Home base. The Air Force put more than twenty million dollars into facilities at the base and the Saylor Creek Bombing Range during 1996 and 1997. In February 1997, the base received a new commissary, and a new base exchange opened in August. Base housing received fifty-six new two-bedroom units, replacing 1940s vintage units. And the enlisted club was remodeled. A new "large aircraft" hangar was built for B-1B Lancer bombers (which had taken the B-52s' place in the composite wing in 1994) and KC-135 tanker maintenance. In addition, the base's 13,500-foot runway—more than two and a half miles long—was designated as an alternate emergency landing site for NASA's space shuttle.

Also in 1997, the Air Force established a "battle lab" at the base to take advantage of the versatility of the composite wing. With a core staff of twenty-five, the battle lab would research better ways to use limited resources. It would research computer-assisted tactics and improved efficiency, including satellite communication that would limit the amount of equipment needed when the wing was deployed. The wing with its variety of planes would test the concepts developed in the lab. The lab was one of six in the country designed to help the Air Force answer the challenges of warfare in the twenty-first century—a strategy think tank.

The future of the Mountain Home base was no longer tied to any range proposal. Some thought the proposed electronic combat range and expanded airspace would become the training ground for the next generation of air superiority fighters—the new F-22. Though talk of base closure still lingered, this specter did not appear to be the sign of catastrophe some feared.

Public hearings on the latest Mountain Home range complex proposal continued during 1996 and 1997. Comments were less vociferous but still overwhelmingly against the range. And in January 1998 the Air Force released a final environmental impact statement on the range proposal. The training complex would be scattered south of the Saylor Creek range along both sides of the Bruneau and Jarbidge rivers. Despite the adamant opposition from BLM and Fish and Game officials, the document included expansion of military airspace over the Little Jacks Creek Wilderness Study Area. The Shoshone-Paiutes objected. Though the practice bombing range had been moved away from any sacred sites, low-level flight over proposed electronic targets still would disrupt religious ceremonies, they said.

Senator Kempthorne said the Air Force had listened to concerns raised by Idaho citizens to protect environmental and cultural resources while meeting training needs. But examples of the voluntary restrictions he cited included not flying over popular recreation areas during Memorial Day, Fourth of July, and Labor Day weekends—a time when pilots typically do not train anyway. The Air Force also promised not to fly over Little Jacks Creek during prime recreation weekends, missing the point that the concern in this area was wildlife, not recreation.

Critics of the proposal said Kempthorne was wrong, and the Air Force had not listened to the public. More than eighty percent of the public comments on the proposal had been critical. In addition, the Owyhee Canyonlands Coalition noted that the public and range opponents were not allowed to participate in a Natural Resource Council, set up by the Air Force to discuss issues relating to the range proposal with Idaho Fish and Game officials. Federal law clearly required that the public be informed of and allowed to participate in such meetings, coalition spokeswoman Lisa Shultz said.[7]

"This is just another example of how the public process that the Air Force, Senator Kempthorne, and others have touted is nothing but a sham," attorney Murray Feldman said.

The proposal did not have the support that Kempthorne seemed to indicate in his public statements. Idaho Fish and Game and BLM officials had serious reservations. The Fish and Game commissioners said they would not support a proposal that did not include at least four provisions:

- Aircraft will avoid critical bighorn sheep lambing areas, as defined by Fish and Game.
- A firefighting plan will include rapid response aerial firefighting capability.
- The Air Force will fund or establish a native plant nursery to provide native seed for range restoration.
- The Air Force will develop and fund a monitoring plan for sensitive wildlife populations that may be affected by range operations.[8]

These four points were not included in the final impact statement. And less than a week before acting-Air Force Secretary F. Whitten Peters was due to make his final decision on the range, Col. Fred Pease, chief of ranges and airspace at the Pentagon, flew out to Idaho to visit with Fish and Game officials in Boise. Before the visit, Pease had said in an interview that the Air Force might not be able to satisfy all of Fish and Game's demands.

Following the meeting, new Fish and Game Director Steve Mealey, who had replaced Jerry Conley, said Pease had assured him that the Air Force wanted to meet the state's requirements. Conley had left to take a job in Missouri for reasons unrelated to the range issue. Mealey was optimistic that the provisions would be included in Peters' decision. Commissioner Richard Meiers was disappointed that the provisions were not already in writing.

"They've told us before they were going to do that, and they haven't," Meiers said. Air Force assurances do not mean anything until they are in writing, he added.[9]

When the Air Force issued its final "record of decision" in March 1998, Fish and Game officials said they were satisfied that changes to the proposal would meet the minimum requirements. Bob DiGrazia of the Foundation for North

American Wild Sheep accused the department of failing its charter for going along with the decision.

Once the Air Force had submitted its final decision, the BLM was to review it and make a recommendation to the secretary of the interior, who would pass a request on to Congress for a land withdrawal. But BLM officials in Idaho would not go along with the proposal unless the Air Force took further steps to "protect public uses plus natural and cultural resources affected by an expanded range"— concerns expressed in January before the final impact statement was released. BLM officials particularly wanted low-level overflights restricted to no lower than five thousand feet over the Bruneau-Jarbidge and Owyhee River canyons during peak recreation periods in April, May, and June—the peak rafting season. And the agency would not support any proposal that included expanding military airspace over the Little Jacks Creek area, state BLM Director Martha Hahn said.[10]

"This is the only (wilderness study area) in southwest Idaho that does not experience low-level training flights," Hahn said in a January 2, 1998, letter to the Air Force. The agency echoed concerns expressed by Fish and Game in 1990 that the area was home to an important herd of California bighorn sheep and critical winter range for mule deer, antelope, and sage grouse. The area provided a valuable comparison to gauge the effects of military operations in other areas. BLM officials also had concerns about the release of chaff and flares, the effects on animal and plant species of concern, and effects on Shoshone-Paiute cultural and sacred sites.[11]

Hahn submitted the summary of her findings and recommendations to BLM Director Patrick Shea in Washington, D.C., who would meet with Air Force officials to try to resolve the issues. It would then be up to Interior Secretary Bruce Babbitt to propose legislation for the withdrawal of public lands for the training complex, Hahn said. She was willing to work with the Air Force on a memorandum of understanding to resolve the issues. And BLM, Interior Department, and Air Force officials began to negotiate a settlement.

Kempthorne was angry. He had thought all major issues already had been resolved. And he was disappointed that the BLM did not bring up the issues at a meeting earlier that year. Kempthorne told the *Mountain Home News* in April that "means could be developed to work around BLM."[12]

In early May 1998, Kempthorne found those means. Rather than waiting for the BLM and Air Force to work out their differences, he inserted an amendment into the 1999 Defense Authorization bill that would set aside the land for the range. He insisted the amendment would not undermine the ongoing negotiations between the Air Force and the BLM, as critics had said. The amendment was merely a "place-holder" until an agreement was hammered out. According to Kempthorne, until the Senate passed the bill, the amendment could be changed to reflect any agreement between the Air Force and the BLM. But the effect was to remove any incentive for Air Force officials to negotiate; they had only to stall long enough for the existing legislation to pass to get the range, critics complained.

"Senator Kempthorne's language takes the BLM out of the loop. It is a backdoor attempt to get the bombing range through, even though it is highly controversial," said Kent Laverty, executive director of the Idaho Wildlife Federation. The amendment undercut the public and congressional process of scrutiny of public land withdrawals normally done in stand-alone legislation, other critics said.[13]

But Kempthorne and BLM officials in Washington said negotiations would continue. In mid-May, ten high-ranking Air Force officers confronted Patrick Shea in Kempthorne's office. With little room to maneuver, Shea gave in to the pressure. Thinking that the agency and the public in Idaho were better served by getting some restrictions on Air Force operations rather than risking all mitigation measures being stripped by Kempthorne's legislation, the BLM gave up the Little Jacks Creek area.

"This is a significant agreement between the BLM and the Air Force which allows us to provide for the needs of Mountain Home's pilots, allows us to protect the environment and allows recreationists to have predictability, which is so very important," Kempthorne said.[14]

Hahn said a negotiated settlement would be better than saying no, which would allow the Air Force to continue current operations without restrictions. But range critics noted that if the current proposal were withdrawn, existing operations still were subject to unsettled legal challenges over changes in operations and airspace that had not been covered in a comprehensive environmental impact statement as required by federal law.

Nevertheless, the agreement would limit low-level flights near the Bruneau-Jarbidge River and Owyhee River canyon areas to one thousand feet above the ground and would require crossing canyons on a perpendicular to shorten the time above the canyons. Flights parallel to canyons would be limited to five thousand feet above the ground, and low-level flights would be limited to that height during peak recreational weekends in April, May, and June. Composite wing exercises also would be limited to two per month during those months. And military airspace would be expanded over Little Jacks Creek, but not below five thousand feet above the ground during April, May, and June.

Bitterly disappointed, critics noted that the restrictions might reduce effects on recreation for a few weekends, but they did nothing for wildlife the rest of the year.

Kempthorne's legislation raised another issue that sparked a national debate. His amendment required the Air Force to compensate Three Creek rancher Bert Brackett for the public land grazing leases he would lose to the range. The Air Force had said all along that it would compensate any rancher affected by the proposal. Compensation in the past for similar changes had been common, but Kempthorne's amendment marked the first time a compensation package had been legislatively mandated. It would have been the first time Congress had ordered a government agency to pay cash to a rancher to give up a lease on land already owned by the government, a precedent that concerned BLM officials. The government paying cash for grazing leases implied a property right that was contrary to a long-standing BLM policy that grazing was a privilege, not a compensable property right.

The Senate had approved the money for the compensation package that ultimately came to nearly one million dollars for Brackett. Senator Craig sat on the appropriations committee that approved the expenditure, and one of his staffers at the time was Jani Brackett, Bert Brackett's daughter. Craig's office and Brackett said she did not work on any issues that involved her father or his grazing leases.

U.S. Sen. Ron Wyden of Oregon was outraged at the idea of paying a rancher for land that already belonged to the government.

"Does this now mean that every federal permittee can expect federal purchase of grazing rights where those are reduced for reasons of resource protection or when policy changes are made through the land management planning process or other public decision-making?" he wrote to Kathleen McGinty. Equally troubling, Wyden wrote, was that Kempthorne's legislation would allow the Air Force to lease that same grazing land back to Brackett.[15]

Eventually the language in the legislation was changed to simply authorize the Air Force to negotiate an agreement with Brackett to provide appropriate consideration. Brackett would lose the leases to 1,252 animal unit months (a unit of grazing management that equals the amount of forage a cow and her calf eat in one month). The compensation package promised Brackett $650,000 in cash plus replacement grazing leases the Air Force would buy from rancher Frank Bachman for $325,000. Brackett would be required to move fences and waterlines and to build a stockwater reservoir.

Brackett's family had owned the Flat Creek Ranch in Three Creek since the 1880s. Over the years the family had assembled a 17,600-acre ranch with attending grazing leases covering 225,000 acres of federal public land stretching from the flanks of the Jarbidge Mountains in northern Nevada to the desert flats along the Bruneau River. Using materials provided by the federal government, the family over forty years had built a fifty-mile stock watering system. The operation allowed the Bracketts to move cattle from the low-lying winter grazing to the higher ground in summer—similar to a natural migration—without trail drives or cattle trucks.

Taking twelve thousand acres out of this massive operation didn't seem like much, but Brackett said it interrupted the continuity of the operation. And he intended to hold the Air Force to its promise to "make him whole." Brackett would get $765,800 worth of compensation in cash and replacement grazing leases for the disruption to his operation and the loss of grazing leases valued at

$173,500, according to a report by Boise appraiser Robert W. Smith of Idaho Land & Appraisal Ltd. Co. The Air Force would buy replacement grazing leases from Bachman valued at $115,800 for $325,000. Neither Brackett nor Bachman owned the land. They were being compensated for the effect the change would have on their operations and for their investments in range developments on public land—fences and water lines. And Brackett would have the first shot at grazing leases from the Air Force on the land for which he had been compensated.

Wyden mustered some opposition. In June 1998, he and Democratic Sens. Harry Reid and Richard Bryan of Nevada, Sen. Daniel Inouye of Hawaii, and others introduced an amendment to strike Kempthorne's amendment, which they complained circumvented the normal process for withdrawing land for military use. The competing amendment sparked a hot debate on the Senate floor.

When pressed on the issue of need, Kempthorne talked at length about the grave dangers in an unstable world and the importance of adequate training for troops that might be asked to fight. In a passionate defense of his amendment, Kempthorne said, "I don't want to be on the side that denied them the opportunity for adequate training." Interior Secretary Babbitt, Katie McGinty, President Clinton, and other officials supported the range because they wanted to ensure that the pilots got adequate training.[16]

Sen. Strom Thurmond stated it even more strongly: "That land withdrawal is necessary to ensure the very realistic military training of the 366th Wing at Mountain Home Air Force Base."[17]

Their comments implied that the Mountain Home wing was less than adequately trained. Yet Air Force officials said that training at Mountain Home was adequate and that the wing was their best-trained unit. Senator Reid in his remarks quoted an Air Force deputy chief of staff who said the range was not strictly necessary for composite force training; the wing already met those needs at the existing Saylor Creek range and ranges in Idaho and Nevada. He quoted Wing Commander Gen. Ken Peck, who said, "We are the most combat capable unit anywhere in the world right now."[18] The proposal was nothing more than a thirty-two million dollar insurance policy for Mountain Home against future base closures, paid for by the taxpayers, Reid said. Adequate training facilities obviously already were available to the wing.

Kempthorne criticized Reid for quoting the 1995 audit report from the Defense Department Inspector General, which said a range in Idaho would be redundant. That report dealt with the state's proposal, which was rejected, "and that is not the proposal before us today," Kempthorne said on the Senate floor. But only a few minutes earlier he had said the project to expand training predated his tenure in the Senate. "It has been around many, many years, but it is time to bring it to a conclusion."[19]

Kempthorne also spoke about the most recent two-and-a-half-year process that included sixteen public hearings in three states, with more than four hundred witnesses and more than a thousand comments. Air Force officials met with the Shoshone-Paiutes twenty-six times, Kempthorne said.

But of all those comments, eighty-six percent opposed the range proposal.

"You can go around and get all the comments you want, if you are going to ignore them. That is what was done here," Reid said.[20]

Groups opposed to the proposal included the Shoshone-Paiute tribes, Taxpayers for Common Sense, The Wilderness Society, the Sierra Club, the Idaho Wildlife Federation, Owyhee Canyonlands Coalition, Foundation for North American Wild Sheep, the U.S. Public Interest Group, the National Wildlife Federation, the Nevada Wildlife Federation, the Idaho Conservation League, Friends of the Earth, the Rural Alliance for Military Accountability, the Oregon Natural Resources Council, the Idaho Whitewater Association, Idaho Rivers United, the Committee for Idaho's High Desert, the Oregon Natural Desert Association, and Friends of the West.

And in southern Idaho, four major newspapers opposed it. "The fact is, the citizens of Idaho oppose this expansion six to one," Reid said.

But Kempthorne's amendment survived Reid's challenge intact. The Senate passed the 1999 Defense Authorization bill later in the summer of 1998, and President Clinton signed it on October 17. Also in the fall of 1998, Kempthorne won his bid for governor of Idaho with little effectual opposition. But many of those who supported his gubernatorial bid said he was wrong about the range.

Chapter 14 / The Final Battle

WHEN PRESIDENT BILL CLINTON signed the 1999 Defense Authorization Bill, which included Sen. Dirk Kempthorne's range withdrawal rider, he undermined any political clout opponents might have had in Congress to combat the range proposal known as Enhanced Training in Idaho. Opponents were unlikely to halt the proposal, but they still had other cards to play that could affect the outcome. State land that would be part of the range had to be leased; issues from U.S. District Court Judge Edward J. Lodge's 1995 ruling on the 1992 environmental impact study and later legal challenges had to be resolved; and grazing permits had to be transferred.

Before construction on the new range could begin, the Air Force needed to lease the state lands that would be part of the range in addition to the federal land already withdrawn. The 12,000-acre practice bombing range would include about 960 acres of state land, and a simulated target site and three remote emitter sites would include another 6 acres. The rest of the range would be on federal land administered by the Bureau of Land Management. The proposed lease came up for approval at the state Land Board's December 15, 1998, meeting—not long after Kempthorne had won the gubernatorial election but before he had taken office. Land Board members agreed to postpone their consideration of the Air Force lease until January. By then Kempthorne had assumed office and as governor headed the Land Board. The delay gave him a key role at the state level in addition to the role he had played in the Senate in pushing the range withdrawal through Congress.

In January 1999, however, the Owyhee Canyonlands Coalition filed a competing application for the same 960 acres, forcing an auction under state law. The coalition submitted a check for $8,548 as an advance payment of the first year's rent—required to qualify as a bidder. In a conflict auction, parties bid what they are willing to pay in addition to the annual rent. The money earned from the lease would go into a school endowment fund.

In the January 12 auction, the Canyonlands Coalition bid $5,000. The Air Force bid $10.

"It's amazing to me that the Air Force can pay out $1 million of taxpayer's money to one well-connected rancher to move some cows around, but only offers Idaho's school children $10," said Roger Singer of the Sierra Club, one of the thirty-two groups in the coalition—the apparent winner.[1] But the Land Board, which has the final say, does not always award the lease to the highest bidder.

"The state land is an integral safety footprint that we need. Going before the Land Board is part of the formal steps," said Col. Billy Richey, representing the

Air Force. "We just need to go ahead with the process and take a look at what we're building for Idaho."[2] And if the Land Board should award the lease to the coalition, Richey said, the Air Force would use whatever appeals were available. He had told the Land Board that the range would not be viable without the state lands and that the range was vital to national security.

The opponents objected.

"There is no statement in the ETI record that the expansion is 'vital' to domestic or other security interests, or that state lands were a keystone aspect of the proposal," Murray Feldman responded.[3] Other Air Force leaders apparently sided with Feldman. In reference to ongoing legal action involving the proposed range, an internal Air Combat Command memo labeled *high importance* had noted, "The Air Force contends that the range represents enhancement, not necessity."[4]

Still, no one was surprised when the Land Board on January 26, 1999, awarded the lease to the Air Force, saying the agency had demonstrated a long-term commitment to the public school endowment fund by leasing the nearby Saylor Creek range since 1962—at an average $0.80 an acre per year. The Air Force would pay considerable more for the new range. The annual payments for the leased state land would average about $6.29 per acre, plus a flat $2,000 for a one-acre site and $250 each for two quarter-acre sites for a total of about $8,548 in annual rent—plus the $10 bid. The lease would have to be renewed in five years.

Soon after the state lease was settled, engineers began pounding survey stakes and bulldozers began rumbling across the sagebrush desert. But the places they were looking at were not all the same ones considered in the environmental impact statement. Range opponents blasted the BLM for approving the construction sites, and they filed an administrative appeal over rights of way the agency granted the Air Force without adequate evaluation. Coalition members wanted their legal challenges resolved before range construction was allowed to start and before proposed airspace expansions took effect.

Judge Lodge's 1995 ruling had left open the question of whether the Air Force should rewrite the 1992 impact statement to include a subsequent range proposal. The Greater Owyhee Legal Defense also had filed suit over the 1994 change in the structure of the composite wing that brought in B-1B Lancer bombers, which are noisier than the B-52s they would replace. GOLD contended the change should have been included in the impact statement along with the range and the composite wing.

In addition to those still unresolved challenges, GOLD had also challenged the adequacy of the 1998 impact statement in U.S. District Court in Boise. That lawsuit was filed in late 1998 just before Clinton signed the appropriations bill that carried Kempthorne's rider. The environmental impact statement on the most recent proposed range was superficial and failed to comply with federal environmental regulations, said Craig Gehrke of the Wilderness Society's Idaho office.

"The Air Force has never done a legal analysis and adequate studies of the effects of flights on wildlife and public uses of the Owyhee canyonlands," Gehrke said.[5] He cited the Air Force's noise analysis that used a model developed to gauge the effects of aircraft noise on city dwellers—it was not appropriate for a wilderness setting. And the 1998 impact statement made assumptions based on the 1992 impact statement. The Air Force should consider changes in airspace and the effects of supersonic flight in a single impact statement along with the composite wing, the addition of the B-1B, and the proposed bombing and electronic combat range complex, Gehrke and GOLD argued.

Air Force lawyer Peter Bogy countered that the effects of noise were covered in the 1992 and the 1998 impact statements. The court also had said the most recent proposal was not related to earlier proposals. It was not necessary for composite wing training and did not need to be considered in the same impact statement, he said.

Some said the most recent suit was filed only to maintain some leverage after Congress had passed the withdrawal bill. But GOLD maintained the impact statement was legally insufficient because it relied on the 1992 impact statement that had been ruled in part inadequate. GOLD also maintained that the legal challenges should be resolved before the Air Force was allowed to go ahead with the new range.

On March 11, 1999, the range opponents asked for an injunction that would have halted construction on the range and would have blocked the Air Force from initiating supersonic operations in the expanded airspace over Little Jacks Creek. The coalition contended that the Air Force should analyze the effects of supersonic flight on bighorn sheep in that area, evaluate the environmental effects of construction, and accurately identify the location of access roads and remote radar sites that would make up part of the range complex. Many of the sites identified in the impact statement had been moved to locations not analyzed—some were off by one hundred feet, others by up to two miles. Until the Air Force had done all these correctly, the judge should not allow construction to begin on the new range, said Laird Lucas, an attorney who had joined Murray Feldman in representing GOLD.

"We don't want to see construction start until this is resolved," Gehrke said.[6]

The Air Force finally took the challenge seriously. Rather than fight it out in court, Air Force officials agreed to halt the construction if the coalition would withdraw its injunction request. The two sides signed an agreement on April 9 that—with a judge's approval—would remain in effect until the U.S. District Court judge ruled on the pending lawsuits. A hearing was set for August 6 in Judge Lynn Winmill's courtroom in Pocatello. The agreement accomplished what the injunction had sought.

Then in May 1999, the Air Force suggested the two sides negotiate a settlement rather than go to court. Questioning the Air Force's motives, some critics said that Air Force officials wanted to show the judge that they had tried to negotiate a settlement in good faith. The Air Force thought negotiation would be good public relations, and would be in line with statements that officials had been responsive to public concerns all during the process to develop the range complex proposal, Col. Richey said later. But probably Air Force leaders had decided not to risk the further delays that litigation would involve, counting instead on being able to negotiate an acceptable settlement quickly. In early July, range opponents agreed to negotiate.

On July 9, after three days of intense negotiation, the Air Force and the Owyhee Canyonlands Coalition hammered out an agreement that finally settled the decade-long challenges to the Air Force's efforts to expand training facilities in southern Idaho. The Air Force for its part got to start—finally—its long-sought practice bombing and electronic combat range. Though conservation and sporting groups did not defeat the range proposal, as many had hoped, they came away with some meaningful concessions.

Opponents had hoped at least to block the airspace expansion over the Little Jacks Creek area. Instead they were forced to choose only seasonal restrictions in one of three areas—Little Jacks Creek, the Bruneau-Jarbidge rivers, and the East and South forks of the Owyhee River. Their attorneys had advised them not to go into the negotiation with a set position but with a statement of what they wanted. They wanted a big chunk of land with a big chunk of quiet, Gehrke said.

But the Air Force would not budge on the question of expanded airspace, and operational requirements would not allow for periods of suspended operations. Officials were willing only to impose increased restrictions on low-level and supersonic operations. And they would allow that only in one of the three areas. They suggested the Bruneau-Jarbidge rivers. Giving up more than that would funnel operations into other parts of the military airspace and increase the effects there.

Herb Meyr pointed out that supersonic operations would not be as frequent over Little Jacks Creek because it was close to the edge of the controlled airspace and to farms, people, and towns. And because of the popularity of the Bruneau-Jarbidge canyons, negotiators figured people would complain more to the Air Force if operations affected recreation there. After much thought, they decided on the canyons of the East and South forks of the Owyhee. New restrictions there would limit low-level flights to no lower than five thousand feet and supersonic flights no lower than fifteen thousand feet above the ground during April, May, and June, except for two one-day exercises each month.

It was hard to swallow the supersonic operations over Little Jacks Creek, but at least the settlement included some additional restrictions, said Meyr, who represented the Idaho Wildlife Federation in the negotiations. He was able to secure an additional $220,000 to monitor the effect of military operations on bighorn sheep. And Air Force officials agreed to seek funding for other wildlife protection and to restoring native vegetation in areas burned by fires caused by training operations. The Air Force also agreed to fund studies of noise levels created by composite wing exercises.

The agreement included moving one of the simulated targets out of sage grouse habitat and restricting the use of some radar sites during times when sage grouse used the areas. One restricted low-level flight area was moved away from a bighorn sheep lambing area, and the 651 square miles of added airspace would not be used until the Federal Aviation Administration approved the addition. And—some thought this was the most important—the Air Force agreed to modify an advisory board to include coalition members. That change gave conservationists and sportsmen a seat at the table with local Air Force and BLM staff and a voice in resolving issues regarding range operations. The Settlement Implementation Group would be jointly chaired by representatives of the 366th Wing at Mountain Home, the BLM, and members of GOLD. To wash down any lingering aftertaste the conservationists might have had, the Air Force agreed to pay GOLD's legal fees.

Not everyone was happy. Some said the negotiators sold out, that they should have gone to court. But during a soul-searching session shortly after the agreement was reached, several negotiators realized they probably got about all they were likely to get—perhaps more than if they had gone to court. The congressional approval of the withdrawal legislation had undermined the coalition's position, said Brian Goller, who represented the Idaho Conservation League in the negotiations. If the coalition had won in court, the judge would not have

Brian Goller, at left, Laird Lucas and Craig Gehrke on a day-hike in Little Jacks Creek area. *Photo by the author.*

been able to overrule Congress, only halted operations while the Air Force completed a new impact statement. But even with a new impact statement, the range opponents might have gotten fewer restrictions. They could no longer count on public support. People were tired of the issue, and were not as likely to show up for another set of hearings on a project Congress had already approved. Overall, the results were probably better than if the lawsuits had gone to court, "given the hand that Governor Dirk Kempthorne dealt us," Gehrke said.[7]

The agreement showed that environmentalists could sit down with the Air Force and negotiate a resolution that was good for the environment and good for the Air Force, Laird Lucas said. It also was the first time Air Force officials had recognized the environmental effects of the proposal and the need to mitigate those effects. Most agreed the agreement would set precedents that would give people in other areas a better chance of negotiating favorable concessions from the military. Though the settlement was less than what the opponents had hoped for, it was better than what the BLM or Idaho Department of Fish and Game were able to accomplish. And with the agreement, the range would be less harmful to wildlife, to recreation, or to ranchers than the Air Force's original 1989 proposed 1.5 million-acre Saylor Creek Bombing Range expansion.

The August 6 hearing date was canceled. Air Force officials and the coalition representatives set out over the next few months to hammer out the details of the agreement. But that took longer than everybody expected. The final agreement was not completed, as expected, sometime in September.

Meanwhile, just when everyone thought the whole thing was over, it became apparent that a BLM grazing permit appeal could unravel the Air Force's plans. The land withdrawal for the range complex hinged on the compensation package for Three Creek rancher Bert Brackett. But one of the key grazing permits that was to be part of the package had been appealed.

On July 9, the same day the Owyhee Canyonlands Coalition and the Air Force reached their agreement, BLM Director Tom Fry asked a federal administrative judge at the Interior Board of Land Appeals to expedite the appeal, which could frustrate the Juniper Butte Range Withdrawal Act.[8] Until the appeal was resolved, the deal could not go through; without the compensation deal, the training complex "will at best be delayed and may be frustrated," Fry said.[9]

Brackett's compensation package included the transfer of a portion of a grazing permit issued in February 1997 to Frank and Cindy Bachman of Bruneau for the Clover Creek Allotment in eastern Owyhee County. But in March of that year, Jon Marvel of Hailey—long a critic of public land grazing and head of the environmental group Idaho Watersheds Project—had appealed the permit. The appeal contended BLM officials had not properly documented their environmental review before issuing the permit.

Under Brackett's compensation agreement, the permit could not legally be transferred until the appeal was resolved. Normally an appeal can take five or more years to be heard, and if the judge upheld it, the transfer could be delayed an additional six months or more while BLM officials completed the review of the allotment and reissued a valid permit. In addition, to transfer only a portion of the permit, the BLM would have to change the allotment boundaries, and that too would be a grazing decision subject to appeal. It could be a long time—five years or more—before Brackett's cows moved from the proposed bombing range to Bachman's allotment.

To resolve the issue, Fry asked the administrative judge hearing the case for an expedited review. "If the appeal were resolved rapidly in the government's favor, Brackett's move to the Bachman tract could be materially advanced and the public interest furthered," Fry wrote.[10] But his intervention turned out to be unnecessary. Bachman's contested permit expired at the end of 1999. The BLM would have time to complete the required environmental analysis of the changes and to issue a new permit before the start of the 2000 grazing season.

With the grazing permit issue resolved, the pieces were falling into place for the Air Force. In mid-November 1999, representatives of the Air Force and the coalition of opponents finally agreed on the terms of the settlement hammered out in July. U.S. District Judge B. Lynn Winmill approved the settlement about a week later. Not long after that, Air Force Lt. Col. Jim Baker from the Mountain Home base told the Twin Falls Rotary Club about the conclusion of the effort and the new training complex the Air Force would soon have. But Baker wondered what the big deal was that led to all the legal wrangling for more than eight years.

"It's just a patch of dirt," he said, showing the Rotarians a slide of the newly started construction.

His attitude explained the frustration opponents of the range had dealt with throughout the struggle. Just when it looked like things were going their way, a new officer would arrive on the scene, who, like Baker, did not seem to be familiar with all the issues. The new Air Force range would affect only 12,000 acres—not the 1.5 million acres once proposed, Baker said. But like others before him, he did not seem to acknowledge that the effects low-flying jets might have on the millions of acres below them would be the same no matter who controlled the land. And that was what all the fuss was about.

People opposed the range because it threatened livelihoods, recreation and wildlife without a demonstrated need. Despite the assertions of Idaho Gov. Dirk Kempthorne and others, Air Force officials could not show they needed it. And the resulting loss of faith was perhaps the greatest intangible cost. Early in the process, Deputy Assistant Secretary of the Air Force Gary Vest had warned of the importance of being able to show a need for the range. While the Air Force was considering Idaho Governor Cecil Andrus' proposed state range, Vest had written to Air Force Gen. John M. Loh of the Air Combat Command that it was "imperative that we have strong, supportable operational requirements identified" in the planning document for the project. The draft environmental impact statement "will encounter difficulty unless backed by operational justification."[11] Rather than a clear statement of need, the Air Force produced bureaucratic and legalistic double-talk. In June 1996, while the most recent range proposal was under consideration, Air Force attorney Peter Bogy said, "It's a need because it's more efficient to have a local training range, but it's not a requirement upon which the composite wing lives or dies."[12]

The lack of a clearly identified need plagued the Air Force proposals during the entire decade-long struggle. Despite repeated requests from Congress, the Air Force never conducted a national assessment of training needs. And the Air Force never showed that the small additional amount of training the proposed range would provide could be done better or cheaper in Idaho. No wonder people objected. Idaho citizens were unwilling to see a spectacular piece of landscape full of wildlife and wonder, and the recreation and livelihoods it provided, degraded by an increase in military operations that already were being accomplished successfully somewhere else. Col. Fred Pease, head of ranges and airspace at the Pentagon, liked to argue that without the new range in Idaho, training activities over someone else's backyard would increase. But he was wrong. There would be no increase; the pilots from Idaho were already training there. At best the Idaho range would mean a little less activity at those ranges.

Throughout the controversy, unclear, conflicting, and false information— outright lies, in some cases—undermined the public's trust. In 1989, Air Force officials tried to use the aging, soon-to-be-retired F-4 Phantom fighters as a lever to get a huge new bombing range and a large expanse of supersonic operating airspace. They later admitted that the new range was not for the F-4s. When that effort failed, officials tried to use a newly formed composite wing as a lever to secure a huge supersonic operating area in southern Idaho. This time, they succeeded. But they still were not being honest. Herb Meyr said the new operating area was meant for the F-22, a fighter then under development that would cruise at supersonic speed. Planning for the F-22 began in the mid-1980s, about the same time Gary Vest suggested Air Force plans for expanded training at Mountain Home Air Force Base look to the needs of the aircraft of the future. Air Force officials admitted the composite wing did not need the range, but the F-22 needs a lot of supersonic airspace.

Rep. Jerry Lewis of California asked when it was okay to lie to Congress. But to whom do people turn when Congress goes along with the lies and refuses to hold the military accountable? The Air Force lied to Congress, to the Base Closure and Realignment Commission, and to the American people, and it all but ignored the courts to get what it wanted. In the end, Congress went along with the Air Force. When the people of southern Idaho tried to hold the military accountable, they were overrun by the influence of power and money. Still, they forced the Air Force to acknowledge the effects of operations on the wildlife and the people on the ground below their operating areas.

The opposition groups forced the issue into public discussion. The result was a scaled-back range proposal with operational restrictions that might protect wildlife and the people who already were using the areas. They forced the Air Force and supporters of the range, who could not get the proposal approved through regular channels, to rely instead on political maneuvering. The legislation that finally created the new range was not the product of an open, collaborative process as U.S. Sen. Mike Crapo claimed. It was not an open, detailed discussion that included the need for and the effects of the range and training operations. It was instead a controversial rider attached to a must-pass military appropriations bill.

The Air Force got the thirty-million-dollar range. But the real cost was far greater; it included the loss of faith in honest government—a government that never required the Air Force to show a need for the range and did not seem accountable to the people of Idaho and the rest of the country.

Epilogue

S ITTING ON A ROCK OUTCROP just off Mud Flat Road one hot summer afternoon, I watched a thunderstorm move up from Nevada. A tall thunderhead grew from nothing to dominate the southwestern sky, and it soon obscured the sun. The rising clouds spread a cast-iron shadow across the desert. Beneath the center of the storm, a gray veil of rain obscured the landscape.

It is hard to predict what changes the Air Force bombing range will bring to this wild place. The fact is that the Air Force already is here, and it is not likely to go away. Restrictions agreed to by Air Force leaders might be an improvement, but it is not clear who will enforce them or how. Whatever the Air Force does, it should be forced to follow the law, and political maneuvering should not be substituted for good environmental and economic analysis and public involvement. As I write this, the Air Force already is considering bringing a wing of F-22s to Idaho. That would fulfill the vague prophecy of Gary Vest some sixteen years earlier.

I heard no planes that day, but on other days I have heard heart-stopping sonic booms and the shattering scream of low-flying jets. Those are foreign sounds here, imposed upon the landscape, unlike the thunder that arises from the desert and rolls away, counting every rock with echoes as it fades. I sought shelter from the heavy raindrops beneath my army-surplus poncho. The wetness leaked through, and cold rivulets ran down between my shoulder blades. Lightning flashed in the darkened afternoon, thunder cracked nearby, as the center of the storm passed a mile or so to the west. The rain left the landscape washed and the air about ten degrees cooler, sharp with the scents of the desert—warm wet rocks, the incense smell of juniper, and the sharp, sweet smell of sagebrush, as pervasive as the solitude.

Appendix

Military expansion proposals spanning the country in the 1990s included:

- Arizona—The Yuma Training Complex proposed to expand training and to develop new electronic target facilities.
- California—The Marine Corps wanted to add 8,300 acres to the China Lake-Chocolate Mountain aerial gunnery range; the Army wanted 310,000 acres to Fort Irwin National Training Center for laser weapons testing; and the Navy 3,320 acres to the China Lake Weapons Center. The Air Force wanted 48,000 acres for the El Centro Parachute Test Range and 54,000 acres for the Titan Missile Range.
- Colorado—The Army wanted to add 5,600 acres to the Pinyon Canyon Maneuver Area near Colorado Springs.
- Hawaii—The Army wanted 38,000 acres for the Schofield Barracks training and firing range.
- Kentucky—The Army wanted to add 33,280 acres at the Fort Campbell Training Range, and 23,700 acres at the Fort Knox Training Range.
- Maine—The Army National Guard wanted 720,000 acres for a tank training area.
- Montana—The Army National Guard wanted to create a 1,533-square-mile tank training range.
- Nevada—The Navy wanted to increase its airspace at Fallon Naval Air Station by 10,000 square miles along with 181,000 acres for a buffer zone around existing ranges and 200,000 acres for electronic training ranges. The Navy also proposed three new military operating areas totaling more than 12,000 square miles, and it wanted to expand its Supersonic Operating Area by 500 square miles. Finally, it proposed 239,000 acres for a land bridge between ranges. The Army National Guard wanted 610,000 acres for a tank range.
- Texas—The Air Force wanted twelve 15-acre parcels for electronic range operations for B-1B and B-52H bombers in Texas or New Mexico.
- Utah—The Air Force wanted 455,000 acres for a training corridor linking the Air Force's Utah Test and Training Range with Nellis Air Force Base in Nevada.
- Virginia—The Army wanted 51,000 acres at the Fort AP Hill Training Range and 2,500 acres at the Fort Eustis Training Range.
- Washington—The Army wanted to add 6,300 acres to its Yakima Firing Center in central Washington.[1]

Notes

Introduction

1. "The Range That Bombed," *Boise Magazine*, (September/October 1990): 30.

Chapter 1. Vision Quest

1. Grace Potorti, director, Rural Alliance for Military Accountability, Reno, Nev., in a letter to the Air Force (July 25, 1996) and at several other times. It is the group's standard position on military expansion in all locations. Potorti also was known by the surname Bukowski.
2. Seth Shulman, *The Threat at Home: Confronting the Toxic Legacy of the U.S. Military* (Boston: Beacon Press, 1992), 149.
3. Dale Ahlquist, Director, National Airspace Coalition, personal communication (email) (April 2, 1997).
4. Ibid.
5. Ibid.
6. Jack Trueblood, personal communication (September 1998).
7. David D. Alt and Donald W. Hyndman, *The Roadside Geology of Idaho* (Missoula, Mont.: Mountain Press Publishing Co., 1989), 34.
8. Bill Bonnichsen, research geologist with Idaho Geological Survey at the University of Idaho, Moscow, Idaho, personal communication (email) (April 16, 1997).
9. Mark Plew, chairman of the Boise State University archeology department, from an abstract of his work from 1976 to 1986 in BSU Archaeological Reports, Boise State University, Boise, Idaho.
10. Betty Derig, *Roadside History of Idaho* (Missoula, Mont.: Mountain Press Publishing Co., 1996), 172.
11. Ibid.
12. Ibid.
13. Ibid.
14. Cort Conley, *Idaho Loners: Hermits, Solitaries, and Individualists* (Cambridge, Idaho: Backeddy Books, 1994), 201-202.
15. Derig, 174.

Chapter 2. Hometown Activist

1. Gary Vest, Deputy Assistant Secretary of the Air Force for Environment, Safety and Health, declined a request to be interviewed for this book.
2. "Bases: A History of Protection by the System," *CQ Almanac* (1988): 441.
3. *The Defense Secretary's Commission on Base Realignment and Closure*, report (December 29, 1988): 6.

4. Jim Courter, Chairman, Base Closure and Realignment Commission, letter to Bill Roberts, managing editor of the *Idaho Statesman* of Boise, Idaho (June 16, 1991). Obtained by the author.
5. U.S. Rep. Jerry Lewis, prepared notes for statement to House Armed Services Committee, Subcommittee on Military Installations and Facilities (February 22, 1989). Obtained by the author.
6. *The Defense Secretary's Commission on Base Realignment and Closure*, 75-76.
7. Air Force memo (February 1989). In author's possession, and referenced in congressional records for hearings before the House Committee on Appropriations' Subcommittee on Military Construction Appropriations (February 23, 1989).
8. *The Defense Secretary's Commission on Base Realignment and Closure*, 16.
9. Congressional records for hearings before the House Committee on Appropriations' Subcommittee on Military Construction Appropriations (February 23, 1989).
10. Lewis.
11. Randy Morris, letter to "Friends of the Owyhee Country" (July 31, 1989). Obtained by author.

Chapter 3. Sensitive Receptors

1. Lee Presley, Glenns Ferry rancher, testimony at public hearing on Air Force range expansion proposal, Glenns Ferry (September 7, 1989).
2. Brian Goller, personal communication (February 1999).
3. Grace Potorti, personal communication (February 1999).
4. Lt. Col. James Cooper, interview with author. Cooper made this comment to the author and in public on several occasions.
5. "The Range That Bombed," *Boise Magazine* (September/October 1990): 32.
6. Exchange with Janet OCrowley, of Picabo, during testimony at public hearing on Air Force range expansion proposal, Twin Falls (September 6, 1989).
7. Randy Morris, of Mountain Home, interview with author (June 1996).
8. Tom Basabe, letter to Capt. Wilfred T. Cassidy, at Mountain Home Air Force Base (June 13, 1989).
9. Gary Vest, Deputy Assistant Secretary of the Air Force for Environment, Safety and Health, news conference, Mountain Home (October 11, 1989).
10. Robert Hall, of Glenns Ferry, testimony at public hearing on Air Force range expansion proposal, Glenns Ferry (September 7, 1989).
11. Idaho Department of Fish and Game, comments on the draft environmental impact statement for the Saylor Creek Bombing Range Expansion Proposal (April 11, 1990).

Chapter 4. Opposing Forces

1. Kelly Everitt, "Panel failed," editorial, *Mountain Home News* (April 25, 1990).
2. "BLM: Thumbs down to Saylor Creek plan," *Mountain Home News* (April 25, 1990).
3. Bob Stevens, of Ketchum, interview with author (February 1997).
4. U.S. Air Force Fact Sheet 96-03, "Sonic Boom" (March 1993).

5. "The Possible Health Impact of Sonic Booms," study by the State of Nevada for the United States Navy, May 1988.
6. Dave Brunner, Boise District manager, Bureau of Land Management, interview with author (September 1996).
7. Rick Atkinson, "Range War in Idaho: Air Force's Vast Expansion Plan is Resisted," *Washington Post* (May 9, 1990).
8. Patrick E. Tyler, "Military Chiefs Detail Plans to Cut Troops, Weapons," *Washington Post* (May 12, 1990).
9. Secretary of Defense Dick Cheney, letter to Rep. Larry Craig (May 18, 1990). Obtained by author.
10. Deputy Secretary of Defense Donald J. Atwood, "Land Acquisition in the United States," memorandum for the Secretaries of the Military Departments (September 13, 1990).
11. Herb Meyr, personal interview, Twin Falls, Idaho (April 1996).

Chapter 5. A Salable Thing

1. Dan Popkey, "Andrus says Saylor Creek fate hangs in balance," *The Idaho Statesman* (June 26, 1990).
2. Cecil D. Andrus, interview with author (August 1997).
3. Chris Black, interview with author (June 4, 1997).
4. Ibid.
5. Cecil D. Andrus, news release (February 8, 1991).
6. Robert F. Door, "It's Time to End Composite Wings," *Air Force Times* (January 16, 1995).
7. Gary D. Vest, Deputy Assistant Secretary of the Air Force for Environment, Safety and Health, letter to Cy Jamison, Director, Bureau of Land Management (May 2, 1991).
8. "Description of the Proposed Action and Alternatives," U.S. Air Force planning document (May 24, 1991).
9. U.S. Sens. Larry Craig and Steve Symms, letter to Hon. James A. Courter, Chairman Defense Base Closure and Realignment Commission (June 3, 1991).
10. Ibid.
11. Report and Recommendation and Order of U.S. Magistrate Judge Mikel H. Williams, U.S. District Court, District of Idaho, Boise, Idaho, Civ. 92-185-S-HLR (October 7, 1994).
12. Ibid.
13. Lt. Gen. Michael A. Nelson, Deputy Chief of Staff for Plans and Operations, U.S. Air Force, memorandum (June 4, 1991).
14. "Let's work together on finding workable range expansion," editorial, *The Idaho Statesman* (June 17, 1991).
15. Delmar D. Vail, Idaho State Director, Bureau of Land Management, Boise, Idaho, letter to Lt. Col. Thomas J. Bartol (August 20, 1991).
16. Gov. Cecil D. Andrus, letter to Cy Jamison, Director, Bureau of Land Management, Washington, D.C. (August 30, 1991).
17. Cy Jamison, Director, Bureau of Land Management, Washington, D.C., letter to Andrus (October 18, 1991).

18. Mike Medberry, Public Lands Director, Idaho Conservation League, letter to Cy Jamison (September 18, 1991).
19. Dave Jett, governor's liaison, letter to Mike Medberry (November 1, 1991).
20. Stephen Stuebner, "Air Force Banks on Its Post-war Luster in Debate Over Expansion," *The Times-News*, Twin Falls, Idaho (November 10, 1991).

Chapter 6. Split Range

1. Gov. Cecil D. Andrus, letter to Norman Guth, Chairman, Idaho Fish and Game Commission (December 19, 1991).
2. Idaho Department of Fish and Game, comments on the Proposals for the Air Force in Idaho Draft Environmental Impact Statement (December 5, 1991): 1-2.
3. Rick Bass, *The Lost Grizzlies: A Search for Survivors in the Wilderness of Colorado* (Boston, New York: Houghton Mifflin Company, 1995), 59.
4. Richard Meiers, Chairman, Idaho Fish and Game Commission, interview with author (April 1992).
5. Stockmen's Training Range Review Committee, the Owyhee Cattlemen's Association, and the "71" Association, written statement (May 5, 1992). Obtained by author.
6. David W. Knotts, of Holland and Hart in Boise, letter to Lt. Col. Thomas Bartol, U.S. Air Force (March 2, 1992).
7. Robert M. Tyler Jr., of Boise, written comments on the environmental impact statement for the Idaho Training Range (July 15, 1992).
8. "Air Force Should Get Hearings out of Closet," *The Times-News*, Twin Falls, Idaho, editorial (September 1, 1993).
9. Brian Goller, of Boise, interview with author (June 1992).
10. Andrew Garber, "Decision Made, Official Says," *The Idaho Statesman* (June 24, 1992).
11. Deputy Assistant Secretary of the Air Force Gary Vest, Air Force memo (November 6, 1992).
12. Gov. Cecil D. Andrus, letter to Gary Vest (July 31, 1991).
13. Ibid.
14. Air Force Secretary Donald B. Rice, letter to R.T. Nahas, of Castro Valley, Calif. (December 3, 1991).

Chapter 7. Range Criticized

1. Gov. Cecil D. Andrus, letter to Interior Secretary Bruce Babbitt (August 25, 1993).
2. Delmar Vail, letter to Brenda Cook, Langley Air Force Base, Va. (October 5, 1993).
3. Delmar Vail, letter to Major Scott Hamer, Langley AFB, Va. (March 30, 1993).
4. Grant L. Petersen, Mountain Home businessman. Prepared statement (June 1993).
5. Jim Baca, interview with author (September 1999).
6. Gov. Cecil D. Andrus, letter to Interior Secretary Bruce Babbitt (August 25, 1993).
7. Ed Marston, "Jim Baca says the Department of Interior in deep trouble," *The High Country News*, Paonia, Colo. (February 21, 1994).

8. Sen. Harry Reid, Congressional Register (September 13, 1993), S11553.
9. Gov. Cecil D. Andrus, comments on Draft Environmental Impact Statement, Boise, Idaho (January 12, 1994).
10. Letter to Interior Secretary Bruce Babbitt, signed by the leaders of fifteen groups opposing the proposed bombing range (October 1, 1993).

Chapter 8. Virtual Wildlife Refuge

1. Phil Lansing of Boise, interview with author (June 1992).
2. Dave Hunter, Idaho State wildlife veterinarian, interview with author, Owyhee County (December 1991).
3. Dwight R. Smith, *The Bighorn Sheep in Idaho: Its Status, Life History and Management* (Idaho Department of Fish and Game, Wildlife Bulletin No. 1, 1954), 20.
4. Paul R. Krausman, letter to Major Robert Kull, Wright Patterson Air Force Base, Ohio (December 19, 1991).
5. Douglas N. Gladwin, declaration to U.S. District Court, Boise, Idaho, Civ. 92-0189 S-BLW (March 15, 1996).
6. V. Geist, letter to the Foundation for North American Wild Sheep (January 6, 1994).
7. Douglas N. Gladwin and Alexie M. McKechnie, "Review of the treatment of predicted impacts to biological resources as stated in the Draft Environmental Impact Statement for the proposed Idaho Training Range" (November 1993): 8.
8. Jim Klott, biologist, Bureau of Land Management, Twin Falls, interview with author (November 1997).

Chapter 9. Mounting Challenges

1. William J. Weida, military economist, "Review of the Socioeconomic Implications of the Proposals in the Draft Environmental Impact Statement on Idaho Training Range, written statement (November 30, 1993).
2. Pritchard H. White, Ph.D., M.E., "Review of Noise Elements of Draft Environmental Impact Statement for Idaho Training Range," written statement, Boise, Idaho (January 14, 1993).
3. Paul Poorman, letter to Brenda Cook, U.S. Air Force, Langley, Va. (January 18, 1994).
4. Fred A. Christensen, letter to Honorable Sheila Widnall, Secretary Air Force (January 20, 1994).
5. Herb Meyr, interview with author (April 1996).
6. Ibid.
7. David Loomis, Combat Zoning: Military Land-Use Planning in Nevada, (Reno, Nev.: University of Nevada Press, 1993), 33.

Chapter 10. Undercurrents

1. Drew DeSilver, "F&G Will Decide Whether or Not to Support Range," *The Times-News*, Twin Falls, Idaho (January 29, 1994).
2. Idaho Department of Fish and Game, news release (February 2, 1994).
3. Idaho Department of Fish and Game, executive summary (January 27, 1994).
4. Gov. Cecil Andrus, remarks to Idaho Press Club, Boise (February 23, 1994).
5. Ed Marston, Publisher, "Jim Baca Says the Department of Interior is in Deep Trouble," High Country News, Paonia, Colo. (February 21, 1994).
6. Bureau of Land Management, Idaho State Office, Background Paper, "Proposed Idaho Training Range National Historic Preservation Act Compliance," Boise, Idaho (May 26, 1994).
7. "War Meetings," The Owyhee Avalanche, Silver City, Id., (from a reprint) (February 17, 1866).
8. Lindsey Manning, personal communication with author (June 1994).

Chapter 11. Legal Challenges

1. Delmar Vail, Idaho State Director BLM, letter to Alton Chavis, Chief Environmental Analysis Branch, Langley Air Force Base, Va. (June 9, 1994).
2. Air Force Secretary Sheila Widnall, letter to Governor Cecil Andrus (October 4, 1994).
3. Gov. Cecil Andrus, interview with author (August 1997).
4. Bob Stevens, interview with author (October 1994).
5. William Brock, "Angry Idahoans Lash Clinton Over Decision," *The Times-News*, Twin Falls, Idaho (October 5, 1994).
6. Report and Recommendation and Order of U.S. Magistrate Judge Mikel H. Williams, U.S. District Court, District of Idaho, Boise, Id., Civ. 92-185-S-HLR (October 7, 1994).
7. Ibid.
8. Murray Feldman, interview with author (February 1995).
9. Ibid.
10. Memorandum, Department of the Air Force (June 5, 1992).
11. Gary Vest, Deputy Assistant Secretary of the Air Force for Environment, Safety and Health, letter to Governor Cecil Andrus (December 31, 1991), quoted in Report and Recommendation and Order of U.S. Magistrate Judge Mikel H. Williams, U.S. District Court, District of Idaho, Boise, Idaho, Civ. 92-185-S-HLR (October 7, 1994): 30.
12. Ronald M. Watson, Chief, Environmental Division, Air National Guard, letter to Eric Christensen attorney for Foundation for North American Wild Sheep (February 18, 1992).
13. Plaintiff's Memorandum of Objections to the Government's Renewed Motion to Dismiss, U.S. District Court, Boise, Idaho, Civ. 92-0189-S-HLR (April 1, 1994).
14. Ibid.
15. Craig Gehrke, interview with author.

Chapter 12. Critical Mass

1. Charles Etlinger, "Aide: Clinton Wanted New Range Plan," *The Idaho Statesman* (November 12, 1994).
2. Gov. Cecil Andrus, interview with author (August 1997).
3. Col. John K. Wilson III, U.S. Air Force, letter to Governor Phil Batt (January 12, 1995).
4. Gov. Phil Batt, letter to Air Force Secretary Sheila Widnall (March 23, 1995).
5. Dave Jett, internal memo (March 3, 1995). Obtained by author.
6. Dave Jett, internal memo (March 8, 1995). Obtained by author.
7. Ibid.
8. U.S. District Judge Edward J. Lodge, Order adopting second report and recommendation, U.S. District Court, Boise, Idaho, Civ. 92-0189-S-HLR, 9 May 1995.
9. Secretary Widnall, letter to Governor Phil Batt (May 9, 1995).
10. Sens. Larry Craig and Dirk Kempthorne, Reps. Mike Crapo and Helen Chenoweth, and Gov. Phil Batt, news release (May 10, 1995).
11. Ibid.
12. Brig. Gen. David J. McCloud, 366th Wing Commander, letter to the editor, *The Idaho Statesman* (December 2, 1993).
13. Gov. Cecil Andrus, interview with author (August 1997).

Chapter 13. One More Time

1. Lasha Johnston, letter to Air Force Secretary Sheila Widnall (June 14, 1995).
2. Brian Goller, "Hard Times on the Owyhee," *The Idaho Conservationist* (Idaho Conservation League Quarterly) (Fall 1998): 5.
3. Ibid.
4. David A. Jett, internal memo (November 30, 1995). Obtained by author.
5. "Environmentalists Try to Stop Plan to Expand Bombing Range," *The New York Times* (August 17, 1997).
6. Settlement Agreement between the Shoshone-Paiute Tribes of the Duck Valley Indian Reservation and the Unites States (August 1996).
7. News release, Owyhee Canyonlands Coalition, July 9, 1998.
8. Idaho Fish and Game Commission, comments on *Enhanced Training in Idaho*, environmental impact statement (July 18, 1997).
9. Fish and Game Commissioner Richard Meiers, interview with author (February 1998).
10. "Summary of Findings and Recommendations," Bureau of Land Management (April 17, 1998).
11. Martha Hahn, Idaho Director, Bureau of Land Management, letter to Air Force Col. Fred Pease (January 2, 1998).
12. "BLM Won't Sign Off on Range Proposal," *Mountain Home News* (April 22, 1998).
13. Kent Laverty, Director Idaho Wildlife Federation, news release (May 14, 1998).
14. Sen. Dirk Kempthorne, news release (May 15, 1998).
15. Sen. Ron Wyden, Oregon, letter to Kathleen A. McGinty, Chair, Council on Environmental Quality, Washington D.C. (June 9, 1998).

16. Sen. Kempthorne, comments on the Senate Floor, congressional record (June 25, 1998), S7070.
17. Sen. Strom Thurmon, comments on the Senate Floor, congressional record (June 25, 1998), S7071.
18. Sen. Harry Reid, comments on the Senate Floor (June 25, 1998), S7068.
19. Kempthorne.
20. Reid.

Chapter 14. The Final Battle

1. Roger Singer, interview with author (January 1999).
2. Col. Billy Richey, Mountain Home Air Force Base, interview with author (January 1999).
3. Murray Feldman, attorney for the Greater Owyhee Legal Defense, Sierra Club news release (January 26, 1999).
4. Col. Stoney P. Chisolm, Air Force memo obtained by author (March 26, 1996).
5. Craig Gehrke, Director, The Wilderness Society's Idaho office, interview with author (March 1999).
6. Ibid.
7. Ibid.
8. Tom Fry, Acting Director, Bureau of Land Management, memorandum to Dale Pontius, Associate Solicitor (July 9, 1999).
9. Ibid.
10. Ibid.
11. Gary Vest, Deputy Assistant Secretary of the Air Force, letter to Gen. John M. Loh, Commander, Air Combat Command in Langley, Va. (April 5, 1993).
12. Steve Stuebner, "Idaho Air Base Guns for More Space, Again," *High Country News*, Paonia, Colo. (June 24, 1996).

Appendix

1. These numbers are from a variety of sources and changed as proposals were dropped or amended.

Index

About the Author

Niels Sparre Nokkentved lives with his wife in Olympia, Washington, where he works as a journalist. His reporting has earned him several awards for investigative journalism. He earned degrees in journalism and environmental studies from Western Washington University in 1988. Born in Copenhagen, Denmark, in 1947, he immigrated with his parents to Canada in 1957, but he spent most of his youth in a northwest Chicago suburb in the 1960s. Niels is an American citizen, and served in the U.S. Navy aboard a destroyer in the Pacific during the Vietnam War.